MW00441095

THE GOLDEN THREAD

THAT WEAVES
A FABRIC
OF REASON

A Brief Treatise
on the Nature of
Existence &Ancient
Alien Theory

By
Dr. David D. Weisher, M.D.

Forewarning
Reading this book may change your life.

Copyright © 2014 Dr. David D. Weisher M.D.

All rights reserved.

ISBN: 0615750575

ISBN 13: 9780615750576

Library of Congress Control Number: 2014901445

David D Weisher MD, St Thomas US Virgin Islands, VI

THE GOLDEN THREAD

There is a golden thread that ties together the many mysteries of the human condition. A common denominator resulting in a new cosmology. Mysteries regarding existence, consciousness, spirituality, UFO encounters, reincarnation, ancient archeological enigmas, ancient alien theory, Biblical as well as ancient Sumerian scripture, and mankind's true evolutionary origin are all part of this great mystery. A golden thread ties together the three basic philosophical questions: Where did we come from? Why are we here? And where are we going? A thread that weaves a fabric of logical conclusion, which is extremely difficult, if not impossible, to deny and certainly can no longer be ignored.

This book is a compilation and digestion of pertinent findings of historical and scientific significance that are presented as relevant points on this golden thread of discovery. Evidence that can no longer be laughed at or suppressed as in the past. Evidence that you may find compelling, if not astonishing. For, as you will see, we are not and have never been alone in the universe.

Did you know that the Great Pyramid is a "perfect" pyramid, and there is evidence that power tools were used? Did you know that the Sumerian *kush* (a sacred measurement given by the "gods"), if used as the length of a pendulum, beats exactly at one second (another measurement given by the "gods")? Did you know that there is evidence that human DNA was tampered and that we are told this by our ancient Sumerian ancestors? Were you aware that the world was almost disintegrated in a massive thermonuclear holocaust in 1968 and that UFO activity was very busy at that time?

If you doubt the findings or conclusions of this manuscript, then you are invited to examine the evidence for yourself. Nonetheless, reading this book may change your life and open your eyes. For all of us have a purpose, great or small, and we are all co-creating with God.

Dr. Weisher explores these six pertinent topics (evolution, ancient scripture, nature of consciousness, reincarnation, archeological enigmas, and the UFO phenomenon) that must be addressed objectively and weaved into a fabric of philosophical conclusion. The author answers the three basic philosophical questions and presents a unified field theory of existence. Dr. Weisher believes that we now have enough scientific data to render a logical conclusion from a neurological/cerebral perspective and presents compelling evidence for the ancient alien theory, reducing it to simple terms so that all can understand. You will also discover that not only do these conclusions not conflict with biblical scripture, but they are supported by it as well.

Imagine that you have been given a serious assignment: to come to terms with reasonable evidence revealing the nature of existence and of the "gods" (provided they exist). Thus you will be researching a unified field theory of philosophy. What topics will need to be considered? Do we have enough data? Will your conclusions be consistent with scientific fact? In order to do this effectively, data must be compiled from multiple sources, including ancient and scientific texts, evolutionary record, and personal testament. Faith and preconceived beliefs will have to be put on the back burner, for we are looking for evidence.

Dr. Weisher concluded that this assignment would be incomplete if it did not contain an examination of a few key topics.

- We must look for evidence regarding the possibility of direct tampering of human evolutionary cerebral development, including the DNA record.
- We must consider evidence regarding the origin of human consciousness and the possibility of reincarnation.
- We must investigate archeological enigmas and discover whether these support any of our previous conclusions.
- Finally, we must not ignore UFO phenomena in this investigation.

This manuscript is a brief treatise based on the extrapolation of facts, not innuendos or touchy-feely expressions of desired conclusions. It is not stuffed with wasted print or bull verbiage—though in some areas the writing is rather technical—so this manuscript can be appreciated by all. It is designed to transfer information and should be considered as a small lamp leading the greater light that some future generation will compile. Dr. Weisher is sure that others will carry the torch in the future.

Thus Dr. Weisher attempts to answer the three basic questions of philosophy in less than two hundred pages. It is what physicists would call a unified field theory of existence—the great secret resolving where we came from, why we are here, and where we are going. A tall order for sure. It is the author's hope and intention that this book will inspire readers to continue the investigation, help weave this fabric of reason, and, perhaps, expand it into a greater tapestry...

About Dr. Weisher

Dr. David Weisher has had a most interesting life. He was diagnosed mentally retarded while in the eighth grade and was placed in special education classes. Little was known about dyslexia in those days, and so it was by great will and effort that he overcame much of his disability and embarrassment. It was a high school principal, Mr. Charles Rathbun, who discovered he was solving all the geometry questions for math students in the ninth grade. He then invited himself to David's parents' home for dinner and refused to leave until David promised to become an engineer. While in college, rather than studying his core subjects in engineering, he was often found in the library learning the Acadian language so he could read ancient Assyrian cuneiform clay tablets. He became an aerospace engineer and helped develop the flight computers for the F-14 Tomcat and B-1 Bomber. He was also a mixed-gas and saturation deep-sea diver, experienced in bell diving as well as underwater welding and plastic explosives for the deep-water oil industries.

While working in the late seventies as an aerospace computer engineer in California, David was informed about a new medical technology called a CAT scan. For the first time, mankind was able to picture a living human brain. Realizing his true passion, human consciousness, he decided to drop engineering and go to medical school to become a neurologist. Everyone, including his fellow workers and even his parents, thought he was nuts. And so he personally financed almost all of his medical education.

Later he graduated from Georgetown University as a neurologist, then he completed a fellowship in electro-diagnostics. While practicing in the Washington, DC, area, his patient clientele was a venerable "who's who" of political power and influence. Dr. Weisher is board certified in neurology, undersea and hyperbaric medicine, and sleep medicine. He has lectured and written several books and papers on topics like hyperbaric medicine, cerebral lateralization, the origin of human consciousness, and the near-death-experience phenomenon.

Although not religious, he has also worked as a medical missionary in the central mountains of Mexico and has traveled throughout the world. He now lives in St. Thomas, in the US Virgin Islands, and practices neurology, hyperbaric medicine, and sleep medicine. Any free time he gets is spent sailing, deep-sea fishing, and diving—when he's not occupied with manuscript research.

DEDICATION

This book is dedicated to the late Walter Funk, who opened my eyes, and to the late Zecharia Sitchin, who opened the eyes of many.

Acknowledgments

Thanks to Dr. Charlotte B. McCutchen, whose encouragement and knowledge was instrumental in the creation of this book.

I would like to thank James and Ginger Martin for their encouragement; Julie Melucci for her editorial assistance; Joanne Bozzuto, Amalia Morrissey, Lynda, and Marshall Hartmann for their encouragement and assistance; and Jo Ann Sickler and Doug Metzger for our many discussions on the topic.

Special thanks to Ryan, whose memory never fails.

In loving memory of Patty Birch and Francesca Greve.

Let him who seeks continue seeking until he finds. When he finds, he will become troubled. When he becomes troubled, he will become astonished, and he will rule over all things.

Jesus, in The Gospel of Thomas
Spoken to Judas Thomas
Discovered in Nag Hammadi,
Egypt, 1945

It riles them to believe that you perceive the web they weave, so keep on thinking free.

The Moody Blues

This manuscript is a compilation of my life's quest: to seek knowledge and understanding more than accolades.

Dr. David D. Weisher

CHAPTER ONE

EVOLUTION
The Index of Suspicion
for the Alien Brain Theory

Something caused our brains to evolve to be much larger and to have more functions than the brains of other mammals.

–Dr. David Haussler, director of the Center for
Biomolecular Science and Engineering,
University of California at Santa Cruz

Ever since I was a child, I have been fascinated by reading old books—the older the better. The idea that someone long ago penned something and I am reading it today has always been a thing of fascination for me. As math and science were not too challenging for me, I spent much of my college years in the library reading old Sumerian texts written on clay tablets thousands of years before the birth of Jesus. I was fascinated by the fact that, like the Bible, the tablets, too, told a flood story; in fact, much of what the Sumerians wrote is similar to the Mosaic text of the first five books of the Bible.

However, it is the science of evolution that has always puzzled me. I believe we can all conclude that evolution is a scientific fact. But the fossil record and the artistic rendering of the "descent of man," from a purely physical perspective, are quite convincing, but from a cerebral point of view, they aren't. Next time you look at the primate evolutionary pictorial chart starting with the chimp and ending with man fully erect, take a careful look at the skull. The posture changes are convincingly linear, but the skull size is not because it suddenly becomes large about 450 thousand years ago. "Something caused our brains to be much larger." That *something* is the point of contention. I am convinced—as are many others—that evolution was not the complete story for mankind's entry upon this planet. There must be another factor in this equation. This brief treatise is about that factor.

This manuscript might seem shocking to some, but I am considered by many to be a very educated person in the fields of science, engineering, and computer logic. I have cheated death many times (mostly underwater), I no longer fear it, and I have lived to be a senior citizen (although I did not expect to). Thus I'm not afraid to divert from my colleagues, as I have a right to an opinion, if supported by facts. So please listen to what I have to say.

DARWIN

Darwin, in his book *Origin of the Species* (1858), did not invent evolution. The evolution concept is as old as the history of civilization itself. It can be found in the ancient cuneiform clay tablets in the ancient royal library at Nineveh (Iraq today), which describe how we evolved from the sea. What Darwin did was to explain how it worked. He provided a mechanism for its function. He revealed how a change in the environment can dictate a physical change in plants and animals.

Anyone who has been to the Galapagos Islands can see the unique contribution that environment makes to the evolutionary process, how the stress of environmental changes can result in evolutionary development, allowing a species to more properly fit that change and survive, and how this environmental change will "select" the best-adapted species and allow their survival. Darwin bore witness to that change by observing the evolutionary changes of the beaks of finches. Each finch on each island has a different type of beak, which allows it to more efficiently sustain itself and survive on its particular islands with its unique plant life. These islands are far apart from each other, so the little finches had to evolve to survive. Well, they did evolve, and they did survive.

The giant tortoises of the Galapagos Islands evolved in a similar manner. Some have very long necks to better eat the plant life found higher up the trees. Marine iguanas evolved a process to spit out salt, allowing them to swim in seawater and survive. Everything from turkeys to tobacco evolved as dictated by the environment. Thus the phrase "survival of the fittest" took hold, and our perception of the world was never the same. Although Darwin was a firm believer in natural evolution as the only process in the development of the human species, few people know that his co-discoverer did not believe this.

DARWIN'S CODISCOVERER

We all know of Charles Darwin as the discoverer of evolution as the mechanism responsible for the origin of all species on earth. However, little is known of his equally brilliant co-discoverer, Alfred Russel Wallace (1823-1913). In 1908, he was awarded the Order of Merit, the highest award given by a British monarch, and was responsible for much of the research that resulted in Darwin's book *Origin of the Species*. Although very humble, allowing Darwin to take credit and world recognition, he took a different twist from their mutual theory. He was convinced that pure evolution alone could not be responsible for all of evolution and believed an extra, spiritual element had a hand from time to time. He believed divine intervention was inserted at three stages of the evolutionary process on earth. First was the creation of life from inorganic matter. Second was the creation of consciousness in animals, and third was the higher cognitive function/intelligence that humans have today. He too recognized the rapid human cerebral development of which there is almost no other explanation in the fossil record other than "divine" manipulation. The evolutionary fossil record was empty then, and it still is today. Thus he believed that "God" used evolution as a tool but also assisted much like a gardener intervenes to help with the objective of the desired plant life. This begs the question: why, for what purpose, and to what end?

"DIVINE" INTERVENTION?

The fact that dinosaurs are extinct is common knowledge, but many still do not understand how their extinction came to be. It is clear that in order for mammals, primates, and, eventually, humans to evolve, these grotesque and violent beasts must be removed; otherwise, small mammalian quadrupeds would be simply food and never evolve to any extent much beyond therapids, small, two-inch, mouse like creatures that lived in the ground and only came out at night to eat insects for fear of being eaten themselves. The key to this great mystery came in 1980, when Nobel Prize-winning physicist Louis Alvarez and his son Walter Alvarez (geologist), discovered that there was a sedimentary layer noted all over the world at the Cretaceous–Paleogene boundary. This is the boundary that marks the end of the dinosaur age and is estimated as being formed 66 million years ago. Only below it do we see dinosaur bones, and above it, in the newer Cretaceous Period, no such bones exist. Thus, this worldwide clay barrier marks the end of the Mesozoic Era and the beginning of the current Cenozoic Era.

With the help of chemists Frank Asaro and Helen Michel, they also discovered that this thin clay boundary (measuring about a hand span) contained a high concentration of the element iridium (hundreds of times that normally found on the earth's crust). Knowing that iridium comes from asteroids, they concluded that a large asteroid struck the earth at this time, causing a global winter, destroying vegetation, and killing off the large and voracious plant- and meat-consuming dinosaurs. Also knowing the amount of iridium in a typical asteroid and the total amount in the boundary, they calculated the size of the asteroid at about 6 miles in diameter. The problem was that no such colossal impact crater existed on earth. The key to this mystery came later when the Chiexulub impact crater was discovered in the Yucatan Peninsula of Mexico. It was difficult to notice because much of the crater is underwater in the Gulf of Mexico. Then, in March 2010, an international panel of scientists endorsed this theory as the cause for the dinosaurs' extinction. Is it possible that some unseen hand had a role in this? Is it possible that this colossal impact was carefully orchestrated by ancient aliens, or was it just serendipity, as most scientists speculate? An asteroid, just the right size and at the right time to create a billion times the energy of the bomb used at Hiroshima or Nagasaki—not too big so as to destroy all life but big enough to do the job—this was the only way for mammals, primates, and eventually humans to evolve.

"Sorry, Michael, there just wasn't enough time."

–Vito Corleone to his son Michael in the movie *Godfather I*

As our knowledge of our world and science grew, there became a serious problem, a problem that refuses to be ignored, although many scientists still do so today. The problem is *time*.

Geological science has been able to place the age of this planet at about 4.6 billion years. This may sound like a long time, but with advancing knowledge of just how complicated we truly are, 4.6 billion years does not seem like enough to get the job done. Paleontologists and evolutionary biologists tell us that the time required to develop a human is only 600 million years. This seems to be true because the geological and fossil record strongly supports this time span. Six hundred million years, from amino acids in shallow pools of water to the human brain, is just not enough time. It became the opinion of several learned scientists that our planet had some outside help from some highly intelligent aliens a long time ago. With the advancing knowledge of our amazingly complicated DNA and physical bodies (especially the brain), many scientists began to question the belief that life started here. There just wasn't enough time.

FRANCIS COMPTON CRICK

Francis Harry Compton Crick (June 8, 1916, to July 28, 2004) was a brilliant biophysicist and neuroscientist who also thought that there was just not enough time. Every high school student studying biology learns the name of Crick. He and James D. Watson defined DNA together and were awarded the Nobel Prize for physiology/medicine in 1962. But what many students did not learn about was his interesting belief in ancient alien theory.

Crick had a particular interest in neurology and human consciousness, and it has always been fascinating to me (having those same interests myself) that one with such an interest and perspective would appreciate the time problem.

Interestingly, Crick was not a creationist. In fact, he often described himself as an agnostic leaning toward atheism. He once said, "Christianity is OK between two consenting adults but should not be taught to children." Also of note, he did not believe in dualistic philosophy—in other words, he did not believe in the soul's or spirit's contribution to human consciousness. He believed that as the brain evolved and became more sophisticated, a pattern of abstract thought was reached that mimicked a soul or spirit without there actually being a soul or spirit.

As I said, Crick and many others had a problem with the time element of evolutionary theory. There just wasn't enough of it to develop life on Earth without a little help.

Crick said in 1993 that he was overly pessimistic about the chance of a purely Earth-based biogenesis for life when he discovered the self-replication protein system for the molecular origin of life (DNA). It was just too complicated and too amazing, and Earth was just too young a planet. Also, all DNA—plant and animal—spirals in the same direction (to the right), indicating a single source. If evolution alone were the cause, then one would expect a fifty-fifty mix of left- and right-oriented spirals. Furthermore, why is there not alternate DNA chemistry? Why is the backbone of all DNA on Earth phosphorus and five-carbon sugar? Why not arsenic or some other kind of sugar? In addition, all of the twenty amino acids, building blocks for proteins, used to sustain life are *left-handed* (organic molecules can be *left-* or *right-handed* in structure depending on their orientation around the carbon atom). If a right-handed amino acid gets into a protein, the entire protein molecule is rendered useless for life formation or for any living organism. All this homogeneity indicates that there is a single source for the beginning of life and that the development of life is *specified* (meaning a definite criterion is required). Therefore, Crick envisioned what he called *directed panspermia*. This is the belief that complex molecules or organisms were somehow transported to Earth by highly evolved beings with greater technology from other planets in the universe. This is opposed to *general panspermia*, in which organisms are transported to other planets purely by chance. (For example, an asteroid striking one planet causes rock material "contaminated" with life-forms or complex genetic molecules to be ejected to space and eventually drift to another planet.)

Crick did not believe that general panspermia could explain biogenesis on this planet. He believed that ancient visitors from many light-years away came here 600 million years ago and sowed the DNA seeds that grew into forests, sea life, insects, and mankind—the *ancient alien* theory. This theory gained further merit when it was quickly discovered that all DNA, plant and animal, on this planet has a right-handed double helix structure known as B-DNA. If panspermia were completely undirected, wouldn't we have closer to a fifty-fifty distribution?

Crick's slight disagreement with Darwin goes beyond the complexities of DNA. In Darwin's day, a "simple cell" was just that—*simple*. However, any serious student of biology will tell you that a "simple cell" is anything but *simple*. It is a fantastic data-storage facility of mind-boggling dimensions coupled with 3-D copying capabilities for repair and duplication, all on a nano scale. Having a foot in both fields, biology and aerospace computer technology, I can state without reservation that this "simple cell" is more complicated than the computers of the F-14 fighter jet and the B-1 bomber combined. In reality, the term "simple cell" is an oxymoron; simple cells do not exist. In addition, there has never been a convincing theory for their primary development. This is the greatest mystery in biological science today. How did anything that complex develop on its own? What happened during the Proterozoic Eon, a million years ago? Most of the chemical reactions taking place in the cell are not natural in the universe and require enzymes to complete. Even more incredible is the fact that there is not much difference between the plant and animal cell structure. Furthermore, there has never been a sensible or convincing theory for *biochemical descent/abiogenesis* (and I know them all). How did these atoms arrange themselves in such a complex and specific way? It is mind boggling, and I can understand Crick's conviction.

None the less, many believe that Crick's theory, *directed panspermia*, has merit and I will tend to agree with them, but as a neurologist, I have been troubled by another problem that is not often talked about. Frankly, it has been pushed under the rug by modern evolutionary science. While our evolutionary process might have had a little help at the beginning, I have been troubled about the end. The rapid cerebral evolutionary development in the last 2 million years is nothing less than astounding. It took hundreds of millions of years to develop *Australopithecus*, a bipedal humanoid that lived about 2.5 million years ago and had an extremely small brain, no more than one-third of the present human cerebrum. *Australopithecus* evolved into *Homo erectus*, with an equally small brain, about 1.8 million years ago.

Then suddenly, as if from nowhere, five hundred thousand years ago, comes *Homo heidelbergensis*, six feet tall and with a brain capacity of about 93 percent of modern *Homo sapiens*. Modern evolutionary science has rejected the "missing link" theory, and I believe this is because they have focused primarily on the physical developments and not the cerebral, despite the fact that cerebral evolutionary development is very *conserved* (i.e., slow). There is a huge gap of paleontological evidence from *Homo erectus* to *heidelbergensis*, a span of only about a million years in which there was an explosion of cerebral evolutionary development. We just don't know how it happened. In an evolutionary time frame, this explosion in cerebral development took place more or less overnight.

Every high school biology student knows that in order for evolution to work, we need population and time. If we have little time, then greater population is needed. If there is a small population, then more time will be required. This can be expressed in this simple formula.

$$T \times P = K$$

T = Time   P = Population

K = the constant number required for evolutionary development (in other words, it does not change in value)

K is always the same number, so if time is large, population must be small, and vice versa.

But between *Homo erectus* and *Homo heidelbergensis*, there is a very poor transitional fossil record (meaning P is a low value) and a very short period of time (T is also a low value). Multiplied together, these two values would *not* result in K. In other words, under those conditions, evolutionary development was not possible. This clearly raises the index of suspicion that something extraevolutionary occurred and gives credence to the theory that our DNA was purposely manipulated by ancient alien visitors 450,000 years ago (Alien Brain Theory).

This begs new questions: For what purpose, and to what end?

What Do We Mean by the Transitional Fossil Record and the "Missing Link"?

The term *missing link* is no longer used among evolutionary biologists. It is a term coined over a hundred years ago and was used to explain the apparent gap in the hominid evolutionary record between the highest living primates to the *Homo sapiens* of today. In 1859, Darwin admitted that the lack of the transitional fossil record was "the most obvious and gravest objection which can be urged against my theory." In defense of Darwin, it should be noted that there was an extreme lack of evidence because geology and paleontology were in their infancy at that time and so had little in the geological record to defend itself.

A *transitional fossil record* is that which exhibits traits common with both an ancestral group and the descendant group. The most famous example is the *Archaeopteryx*, discovered in Germany in 1861. *Archaeopteryx* is considered the transition animal that bridges land-roving dinosaurs to the birds. It lived about 150 million years ago, stood about half a meter in length, and had features clearly identified with the dinosaurs, such as sharp teeth, a long vertebra tail, three fingers, and a long hyperextensible second toe. Of course, it also had feathers and wings like modern birds. (The fact that this was discovered just two years after Darwin's statement, despite the fact that it was over 150 million years old, is rather provocative.) For humans, there is really nothing in the transitional fossil record (no "bridges") to describe the tremendous, rapid cerebral development our ancestors underwent.

In *New Scientist* (February 27, 2008), D. Prothero said that the term "missing link" is outdated and no longer used by evolutionary biologists, suggesting that the transitional record is now complete. As a neurologist, I very much beg to differ. Something very rapid and profound occurred four hundred thousand to five hundred thousand years ago that affected our brains and, to a lesser degree, our appearance. There is a growing belief that evolutionary scientists should acknowledge the complete and utter lack of a fossil record for the human brain and should bring back the term "missing link" and apply it as it was meant to be: as an unexplained gap in the hominid cerebral evolutionary record. Also, they should acknowledge the possibility of animal gene manipulation by ancient alien visitors (Alien Brain Theory) as explanation for that gap.

THERE IS NO MOUNTAIN
Why Recent
Is So Relevant

Clearly, within the last few hundred thousand or even million years, an explosion of cerebral development took place. Having happened so recently, the fossil evidence should be great and abundantly clear. There should be a mountain of evidence, but this mountain is not even a molehill. In fact, it is flat as a pancake. Basically, it doesn't exist.

We can clearly plot the course of primate evolution from *Plesiadapis*, a small quadruped that survived the dinosaur extinction (65 million years ago), to *Proconsul africanus* (35 million years ago), which is believed to be an ancestor of apes and other similar primates. But for the recent explosion in cerebral evolutionary development, we have next to nothing. Something outside the normal evolutionary process clearly must have happened to rapidly develop the human brain in such a short time. Nature is interested in survival, not music and art appreciation. Something happened that was not completely natural to this planet.

The only biped mentioned between *Homo erectus* and *Homo sapiens* is a highly theoretical being called *Homo antecessor*, who is claimed to be the common ancestor to *Homo sapiens* and the Neanderthal. (Remember, Neanderthal is not a progenitor of humans, only an evolutionary branch.) But there is no significant body of evidence, despite the fact that we have been digging for over a hundred years. Nonetheless, we gave *Homo antecessor* a name, so we believe he must have existed, somewhere. Thus *Homo antecessor* becomes the "party line."

Any detective will tell you that a new crime scene is much better than an old one. So one would expect a mountain of living and extinct evidence for the most recent evolutionary changes, and, as we go back in time, less evidence would be available. One would expect similar primates of different species living among the most evolved. One would also expect significant, hard evidence in the dirt revealing extinct primates, because the "crime scene" is so recent.

However, this is not the story on planet Earth. There should be a mountain of not only living evidence but also definitive, in-the-dirt evidence from the recent past. But there is none. For over a hundred years, we have been looking for this missing link, and there is no link to be found. We have been looking for this mountain, and it just isn't there. Every few years we get a discovery from Africa of a possible missing link, but I have always been disappointed. The evidence should be everywhere, and it seems to be nowhere. We have to face the fact: *there just ain't no mountain.*

LET'S TAKE
ANOTHER PERSPECTIVE

Let us just take a firm grip on the obvious for a moment and examine this enigma from another perspective.

Our planet is completely dominated by humans; there is no other species that even comes close to what we have accomplished. All other animals live on an encounter basis. Evolution has given each species its various drives, increasing chances of survival, but understanding these drives is not important.

For example, the mother rat licks the scrotum of the infant male, which in turn stimulates the production of testosterone (helping to form a male brain) and as-sures a male-behavior response to females after puberty. This helps assure further generations of rats. But clearly the mother rat is not thinking of this while she's licking the balls of her infant males. She has been given a drive by nature to lick the balls; that's all. The genes giving the mother rat this drive survived because the behavior fosters production. Nature is interested in results, not understanding.

On the other hand, humans are completely differ-ent. We plan, discover, and change things like no other species. We humans have extracted minerals from the earth and built roads and houses and filled them with automobiles and beautiful things—devices like televi-sions, radios, and computers and vibrant cultures of en-tertainment, art, music, comedy, history, current events,

and science. Our brains have "discovered" human ethics and a concept for justice, a sense of right and wrong.
We have exploited plant and animal life and developed wonderful culinary skills. We have conquered flight and have exceeded the speed of sound. We have grown to understand fundamental laws of our universe in regards to the atom, particle physics, and the relevance of time and its relation to speed, space, and gravity. We have even exceeded the bonds of Earth and have gone to the moon, not once but many times.

Let's not fool ourselves. No other primate even comes close. We are very, very different.

Racial Influences in Explaining Away the Cerebral Gap

Not too long ago, evolutionary scientists from all over the world would explain this cerebral gap by staging various races on the scale of evolution. If Caucasians were plotting the scale, they would put their race at the top, of course. East Asians would put theirs at the top as well. There has been an argument for every race belonging at the bottom or the top of the evolutionary pile.

We now know that this is all bullshit, for all the races clearly have the same brain capability. We can no longer use race as an excuse to explain away this cerebral evolutionary gap. All of us, regardless of race, are simply the same. It took no longer to develop the German than it did to develop the Mayan or the

Australian aborigine. We are in the same boat. So if we are all at the same stage on the evolutionary tree, is it possible that human DNA was manipulated from time to time by highly technically advanced aliens to achieve some purpose?

THE TOBA CATASTROPHE
The Manipulation
of Chromosome Seven?

Ask any geologist about this theory, and be prepared for something cataclysmic. There is compelling geological evidence that there was a supervolcanic eruption at Toba Lake in Sumatra, Indonesia—about 70,000 years ago. This eruption coincided with the beginning of the last glacial period and ushered in a devastating volcanic winter and essentially created the Toba Lake. This resulted in what scientists term a "bottleneck" in the human evolutionary process. As suggested by Rampino and Ambrose, the human population was so devastated that only three thousand to ten thousand were left on Earth.

Ever since the hyoid bone was discovered in the Neanderthal man, theories regarding the beginning of speech and language have been all over the chart, from four hundred thousand years ago until fifty thousand years ago. It is interesting to note that evolutionary biologists, such as Henry Gee (editor of the scientific journal *Nature*) and Salikoko (of the University of Chicago), are beginning to believe that human language really started only about fifty thousand years ago. Is it possible that this cataclysmic volcanic catastrophe was a purposefully orchestrated event by aliens for the purpose of grafting an alteration of the FOX P2 (DNA gene responsible for language) into the remaining humans' genome (chromosome seven) and eliminating the other bipeds? Remember, there is no evolutionary fossil record for language development. The human laryngeal apparatus is so different from that of any other primate or biped that one can consider a *language* big bang theory in addition to the *brain* big bang theory.

In order to obtain speech, a primate will need not only a much longer oral cavity and a lower larynx but, in addition, the cerebral skills to understand and express speech as found in the human left hemisphere, Wernicke's and Broca's areas, respectively. Thus, speech and language is amazingly complex, involving anatomic and cerebral development, and apparently it blossomed on the evolutionary tree a relatively short time ago. This raises the index of suspicion for a highly intelligent alien element, coaxing and nurturing our evolution, advancing the DNA of a few and eliminating the others. Also remember, if the earth is the petri dish, they could not just use another "dish." The only other option would be that all other species must be killed off and then the change could begin. This sounds cold and cruel, but these aliens are gods and can do pretty much what they want to accomplish their goal—but what is their goal, what is their purpose, and to what end?

EVIDENCE FOR
ANCIENT ANIMAL
DNA MANIPULATION

Most every biology student knows that the chimpanzee is our closest living primate relative, with DNA that is about 98 percent identical to ours. Physically, we can recognize the similarity. They have similar livers, kidneys, and vertebra. They have ten toes and ten fingers. They have spleens and similar digestive systems, and they're semi-bipedal.

However—and this is a *big* however—the difference between our physical knowledge and our cerebral knowledge is very great. We know a great deal about the embryological evolution of the physical body and are now only beginning to understand the evolution of cerebral embryology. And what we are learning is astonishing.

The human genome contains approximately thirty thousand specific coding sequences for genes. This is roughly only 2 percent of our entire DNA. Is the other 98 percent only junk DNA left over from unsuccessful past evolutionary changes? Is it just there for no special purpose, like nipples on a man? Or is there something more for us to learn from this apparent "junk" DNA?

LET'S FIRST EXAMINE
HOW DNA WORKS

Any high school biology student knows how genes work. (I'm going to be a bit technical here, but please

bear with me.) There are four molecular nucleotides—cytosine, thymine, adenine, and guanine.
These nucleotides form long chains, linked together
by phosphate. The structure of DNA consists of two
of these chains of nucleotides wrapped around each
other, much like two snakes having sex. This binding
is called a *double helix*. A nucleotide on one chain connects to a nucleotide on the other chain, and these two
connected nucleotides are called a *base pair*. A base
pair is, therefore, any two combinations of nucleotides: C-T, C-G, A-T, and so on. Three of these bonds or
steps form a code, and the proper sequencings of these
codes form one of the twenty amino acids that are the
building blocks of life.

The double-helix DNA in the cell nucleus becomes
unraveled at a certain point, and the two chains are no
longer connected. The base pairs are therefore divided,
and these exposed half steps form another coded molecule, called RNA. This is called *messenger* RNA. The
RNA is then transported from the nucleus to the cell
body (cytoplasm), where it is transcribed into forming
the specific protein required at that time.

In summary, the process is DNA → messenger RNA
→ protein. Years ago, it was thought (in our naiveté) that
this was all that was needed for us to make a human.
Combine and activate the thirty thousand genes in the
proper sequence, and—puff! We have a human.

But there appears to be more—much more. Especially
when we concern ourselves with the human brain.

WHAT DOES THIS HAVE TO DO WITH ALIEN BRAIN THEORY?
Scientific Evidence for
Ancient DNA Manipulation

In August 2006, the prestigious journal *Nature* reported a most interesting discovery. The discovery was presented by Dr. David Haussler, director of the Center for Biomolecular Science and Engineering at the University California, Santa Cruz, and by Katherine Pollard, PhD, assistant professor of statistics at UC Davis. This was an ambitious study and had the cooperation of science centers in France and Belgium. What they discovered was a unique set of genes that apparently function as instruction for cerebral embryological development that occurs during human conception in the womb and early child development—in other words, the very chemical commands needed for developing a brain. They did this by comparing the DNA of primates, humans, and other vertebrates and finding unique areas that appear to deal with brain cortical development.

The cognitive (or thinking) portion of the brain is in the six cellular layers of the cerebral cortex. This is the outer shell of the brain, nearest the skull. What we know of the embryology of the cerebral cortex is that during development of the early brain, special nerve cells are formed that are instrumental in guiding the neurological wiring of the developing brain. These nerve cells are known as *Cajal-Retzius neurons* and seem to act much like a construction foreman at a building site. These neurons release an important protein called *reelin*, which in turn helps build the scaffolding that gives guidance for all the connecting neurons—kind of like making sure all the red wires go to the red outlets and the blue ones to the blue outlets. Obviously, this is a very important function, for if something goes wrong, the end result could be anything from seizure to personality disorder or even no cognitive function at all. It is for this reason that this portion of the DNA is very *conservative*—i.e., it changes very little for thousands of years. Nature wants survival, not epilepsy.

Another interesting thing about these genes is that they do not produce any protein. Rather than operate in the traditional gene fashion (DNA → messenger RNA → protein), these genes simple go from DNA to RNA and stop there. The RNA molecule then falls back on itself and forms a *functional molecule.* That is, this molecule has a particular function, much like a protein would have, but it is not a protein. These "genes" are often referred to as *functional RNA genes.*

Now if that is not interesting enough, we have also discovered that these genes are very *conserved*—i.e., resistant to change. Nature does not want to take too many chances with cerebral development. She wants to be sure that the boys are attracted to the girls and the girls to the boys. Her goal is procreation, and this is very much dependent on human behavior, which in turn is dependent on DNA and "proper" cerebral development. Mother Nature may favor change, but in regards to the brain, the evidence is that she takes her time. Cerebral evolutionary development is conserved.

WE ARE MORE THAN JUST DIFFERENT FROM EARLY PRIMATES
The Manipulation of Chromosome Twenty

Now for something even more interesting: Human genes are very different from those of our nearest primate relative, the chimpanzee. Our physical bodies may have similar genes, but our brains are a quantum leap beyond our primate ancestors, and mainstream science has ignored this enigma for too long. We humans are not simply just another step from the primates.

It appears that somehow the whole process of this relatively recent cerebral evolutionary development was extremely accelerated. It is for this reason that these genes are now called HAR—human accelerated regions. The HAR gene currently of most interest is the HAR-1F gene. This gene is on the long arm of chromosome twenty, has a sequence of eighteen base pairs, and seems to be partly responsible for the Cajal-Retzius neurons described previously. This part of cerebral development is responsible for intelligence. This is out of a total of 118 base pairs. The difference from the chimp to the human is eighteen base pair genes, while the difference from the chicken to the chimp is only two base pairs.

Thus, the areas responsible for human cerebral development appear to be nine times more aberrant (different) from the chimp to the human than from the chicken to the chimp. We estimate the evolutionary time line from the chicken to the chimpanzee to be about 300 million years. If we take a linear view (and there is little reason why we should not), the time line for human cerebral evolutionary development from chimpanzee to human should be nine times as long as that from chicken to chimp—nine times 300 million, which is 2.76 billion years.

But humans appeared only 7 million years after chimps. Thus, 2.67 billion years of evolution took place in the span of only 7 million years. In fact, considering the lack of a transitional fossil record, it might be a few *thousand* years. It was, in short, an explosive flash of cerebral development that cannot be explained by evolution alone. And that raises the index of suspicion for animal gene manipulation by ancient alien visitors four hundred fifty thousand years ago. This is Alien Brain Theory, and it simply makes sense.

It becomes even more incredible when we add the paleontological evidence to the pot. Let's start with the chimpanzee (about 7 million years ago). We can follow the cerebral evolution forward to *Ardipithecus* (about 4 million years ago—interestingly, *Ardipithecus* has a smaller brain than the chimpanzee before it). Then we get to *Australopithecus* (3.6 million years ago). This was a bipedal with humanlike footprints. But there was still no significant change in brain capacity.

Then we come to *Homo habilis* (about 2.5 million years ago). Some stones were found nearby that might have served as tools. This should not be too surprising, because even chimps are known to use a straw as a tool to eat ants from an anthill. Nonetheless, there was no significant change in cerebral capacity; 2.5 million years ago, humanoids had one-third of a human brain. With *Homo erectus* (1.8 million years ago), there was still no real change in brain capacity. I think you can see where I'm going with this.

Then suddenly, as if from nowhere, comes *Homo heidelbergenis* (about 500,000 years ago) with a brain about 93 percent of *Homo sapiens'*. As if by magic, poof, suddenly we have an animal that also has human cerebral capacity. In addition, there is little or no paleontological evidence that evolution ever took part in this change. But there should be: the evolutionary "crime scene" is only a few million years old.

So we now understand, no question about it, that something drastic happened 500,000 years ago, creating *heidelbergensis*, and then 200,000 years ago, leading to *Homo sapiens*. Something we know nothing about caused a huge gene manipulation that resulted in us, humans. It was not evolution alone but external influence, the insertion of alien DNA, triggering the fruition of our *alien brains*. This was not just a chance change but rather a genetic explosion resulting from DNA manipulation by ancient alien visitors. And, oh, how we took over the earth. We clearly dominated this planet with our new alien brains.

WHY DOES MODERN ACADEMIC SCIENCE IGNORE THE TIME PROBLEM?

There is no evolutionary scientific explanation for this cerebral explosion other than external gene manipulation. Yet modern evolutionary science refuses to examine the problem logically.

It's as if one day you examine your bank account and find $10,000. No surprise; you've been saving a little money each pay period. Then a month later, you find a million dollars in the same account. Obviously, something happened. Maybe a bank employee or a rich uncle did you a favor, but something did happen. It did not just grow there. Money doesn't grow in banks any more than it does on trees.

That's what the human fossil record is like, yet modern scientists are mum on the topic. Why? Because modern scientists almost invariably work for someone else—the government, via organizations such as the NIH (National Institute of Health) or other academic institutions dependent on federal funds. This bureaucratic chain of command often results in a "party line" that is not to be crossed.

I have known good friends (brilliant cerebral scientists) at the NIH forced to resign their positions because they accepted the possibility that the origin of human consciousness does not lie solely in the brain and that most likely it arose from a spirit or some axtracerebral source. Their employers thought that this was too close to religion and feared a reduction in funding. Thus, the "party line" becomes a powerful influence on the writings and teaching within the scientific community.

Francis Crick, having won the Nobel Prize and being a natural freethinker, did not have such a problem, and he clearly stated his belief in ancient alien theory. No one would dare ridicule him. Is it also possible it has something to do with his being English, so he was more at liberty with his thoughts in his country, Great Britton, than our own scientists here in America? American academics will deny it, but I disagree. I know too much.

It is so ironic that the only American academic institution apparently unafraid of such academic fascism is the University of Virginia, which now boldly proclaims reincarnation as a fact. The reason for this contrast is simple: its mission statement, written by the university's founder, Thomas Jefferson, makes it very clear that truth will always be the objective. He knew, as clearly stated in his many letters, that governments gravitate to a form of social and academic totalitarianism. He wanted his university not to be subjected to this dictatorial force, and I believe, in this way, he succeeded. Our leaders would do well to again read letters and manuscripts penned by great men like him.

WHAT ABOUT OUR
ANCIENT ANCESTORS?

We now understand that physical, biochemical, and paleontological evidence clearly states that something profound happened long ago, but it stops there. We now know that the fossil record does not give us an evolutionary explanation. It simply says something happened; it does not explain who or why.

It just might not all be serendipity. Is it possible that for millions of years, this planetary biosphere has been manipulated by highly sophisticated aliens much like a gardener hoes, weeds, plants, and cultivates his garden, and, if so, does further evidence exist?

Perhaps we should go to our ancestors and see what they have to say. After all, weren't they closer to the "crime scene," from a time perspective? As incredible as it may seem, they do have an explanation for what transpired.

As in any crime, we must look for witnesses. So if the fossil record is not helpful, maybe the ancient literary record will be of assistance. It is not without relevance that our ancient forefathers did record that something profound occurred at that same time frame as the leap in human cognition. Common sense tells us that this is not just coincidence. When the fossil record states that something profound occurred to our ancestral bipedal brain 450,000 years ago, and the Sumerian texts also say something profound happened to our ancestors' brains at that same time, then I think we should listen to their story.

CHAPTER **TWO**

MYTHOLOGY
Clues to Reality

**Myths are clues to the spiritual potentialities
of the human life.**
–Joseph Campbell to Bill Moyers (1988)

WHY MYTHOLOGY MATTERS
The Amazing Story of Heinrich Schliemann

One of the most incredible—and *true*—stories that deals with mythology and archaeology is the story of Heinrich Schliemann, his wife, Sophia, and the famous city of Troy. Homer poetically wrote of Troy hundreds of years before Alexander the Great. It was said that Alexander always kept Homer's books (*The Iliad* and *The Odyssey*) at his bedside all through his campaigns of world conquest.

Heinrich Schliemann was born in Germany in 1822. His father was a clergyman, which meant that he was poor but well educated. Heinrich grew up hearing the story of Homer's Troy from his father. The story, being a favorite of both of them, was read over and over again until Heinrich almost knew it by heart.

Perhaps you remember the story of Troy. It was a great city on the Mediterranean coast of Asia Minor, what is today Turkey. It commanded the entrance into the Black Sea and so was of vital economic and strategic importance. King Priam of Troy had a son named Paris, who was something of a dandy, a handsome young lad who had more interest in women and fun than in politics. While on a trip to Greece, he ran off with the wife (Helen) of the king of Sparta (Menelaus) and took her across the sea to Troy. Helen was the most beautiful woman in the world, and Menelaus wanted to retrieve her and punish Troy for the crime.

So he told his brother Agamemnon, king of Mycenae (and all of Greece). They planned to attack Troy by "launching a thousand ships." All the great men of Greece were with the king in this endeavor, including Ajax, Achilles, Hector, and Odysseus. *The Iliad* ends when Hector is killed in the attempt to take Troy. *The Odyssey* continues the story with the tale of the Trojan horse, which finally allowed the city to be breached, resulting in the destruction of Troy. The book then continues with the ten-year adventure of Odysseus afterward.

Almost all the scholars, historians, and other experts of the nineteenth century were convinced that the story of Homer's Troy was nothing but a myth and that the city of Troy never existed. As for Heinrich Schliemann, in 1829, when he was only seven years old, he was personally convinced that the story was true and that someday he would go to Turkey and dig up the city of Troy. No doubt many thought this was merely a cute display of passion, but Heinrich never forgot it. Later in life, he set out with his new, young wife, Sophia, to do just that.

Heinrich was a very bright man and was fluent in fifteen languages by the time he was thirty-three years old. He also became a US citizen and amassed a fortune in the import/export business. A multimillionaire at age forty-six, in 1868, he and his wife set out to find Troy.

All the experts, historians, and scholars of the time thought that Troy, if it existed at all, was near the small village of Bunarbashi, located on the coast of Asia Minor. This was primarily because Homer wrote that there were two springs at the site, one hot and the other cold. Bunarbashi had a total of thirty-four springs. However, all of them were cold. Nonetheless, the experts considered the matter no further. Bunarbashi was their site.

Heinrich, on the other hand, considered Homer in more detail. He knew that the Greeks made up to three trips a day to Troy from the coast. Considering Bunarbashi is eight miles from the sea, this site was unlikely. Also, Achilles had chased Hector three times around the city before killing him, but Bunarbashi is surrounded by rocky terrain, making that chase impossible.

Heinrich did something that archeologists rarely do even to this day: he consulted a geologist, who informed him that it is not unusual for springs to dry up after three thousand years, whether hot or cold. Neither Alexander nor King Xerxes visited Bunarbashi. They did, however, visit and pay respects at another site not too far away.

This is the small town of Hissarlik, which is about three miles from the coast. Both of them offered sacrifices at the site, and in fact Xerxes sacrificed a thousand oxen there. Alexander ran around the "city" with his best friend and possible lover, Hephaestion, and retrieved a shield from the site, which he carried with him throughout his campaign. There was also an ancient Roman city there called Novum Ilium, which dated back to 300 BC. Also of interest is that Hissarlik means "palace."

Schliemann believed that this was the site and was willing to stake his sweat and fortune on it. He began digging in 1871 and worked day and night until 1873. In May of that year, after discovering a host of archeological findings, he planned to quit on June 15. Then, as incredible as it may seem, on June 14 he discovered a great amount of gold jewelry, including sixty golden earrings, 8,700 gold rings, gold bars, goblets, and much more. He told his wife to tell the workers to take the day off because it was his birthday. "But it's not your birthday," she said. He said, "Tell them anyway." In this way, he prevented a violent mutiny and preserved the treasure.

To make a long story short, Schliemann did find the city of Troy, and it was, indeed, at Hissarlik. Scholars and archeologists, including Rudolph Virchow and Friedrich Dorpfeld, later confirmed it at level site 7-A. Troy was real. And if Troy was real, then most likely Homer's story was real. Homer's epic was not just an entertaining story but may have been based on real people and real events.

And if Homer's stories are real, there may be some real-life basis for other ancient writings, some much older than Homer.

They Drew and Wrote Because It Was Important to Them

We have all seen the drawings of figures and scenes painted by our early ancestors thousands of years before the pyramids were built. Some of these drawings are estimated to be twenty thousand years old. These paintings were not fairy tales or little novelettes but, rather, events important to them. Things they saw. Things they believed. Things they felt important enough to leave to their children's children.

I believe also that much the same can be said about our ancient Sumerian ancestors, from Mesopotamia (now Iraq), who wrote on clay tablets four thousand years before the Christian era. Their type of writing is called *cuneiform* (wedgelike), and much of it comprises accounting records and items loaned or stored. Occasionally, one finds accounts of kingly conquest and prisoners captured or items confiscated. Nonetheless, nothing was drawn or written down unless it was relevant. There was no printing press to turn out hundreds of thousands of copies for profit and publication at the time.

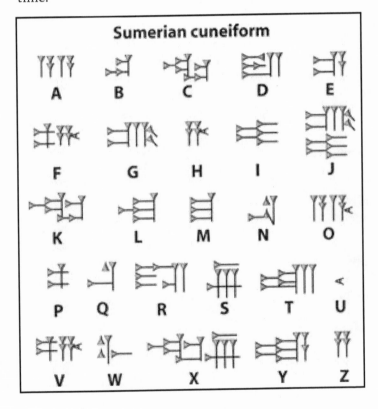

Sumerian cuneiform

A B C D E

F G H I J

K L M N O

P Q R S T U

V W X Y Z

What About Ancient Sumer?

Almost every history scholar I know has been fascinated by Babylon and ancient Sumerian history. Everything in human culture today started in Sumer. As if from nowhere, six thousand years ago on the banks of the Persian Sea in what is today Iraq, appeared a tremendous culture. Suddenly we had laws, multistory housing, mathematics, writing, medicine, music, astronomy, merchants, temples, palaces, streets, and government in the form of kings, courts, and judges.

When I was in college learning to become an electrical engineer, I was supposed to be studying calculus and Boolean algebra. Instead, I was constantly going to the library to get any information I could on ancient Sumer. I was trying to learn the Akkadian language so I could read the cuneiform writing for myself. I read the Gilgamesh epic and the Sumerian account of the Great Flood—anything I could get my hands on. I was super fascinated because I believed that something that old and that advanced must have something important to tell us about our past, some great truth that has been lost to us through years of antiquity.

I know I was right, because they did have something to tell. And what a story it is...

In the first chapter, we were confronted with the fact that something incredible happened around half a millennium before recorded history, something that drastically changed the course of our planet's biological history. In a cosmic flash, as if from nowhere, came the human brain and its 100 billion neurons (a baboon brain has about 14 billion neurons). How did this happen? There is no conclusive biological explanation, and the rationales given to us are just that—*rationales*, without any convincing facts or scientific explanation. The fact is that we just don't know what happened.

But the ancient Sumerians believed they did know, and they wrote it down for all of us to read. It is interesting to note that the story actually begins a long time ago before mankind came to being, and it comes to us from the ancient clay cuneiform manuscript known as Enuma Elish (written four thousand years ago and no doubt copied from older text), meaning *From on high*, the first three words from that great Sumerian text. It is a story of war and rebellion.

THE GREAT REBELLION LONG BEFORE MANKIND

I will give you a short, paraphrased version of this story of a great rebellion among the gods that predates the formation of mankind. As the story goes, the great female goddess Tiamat marries Kingu. This elevates Kingu to great prominence and temptation. He decides to take total command all for himself and makes war with another god, Enlil, and his second-in-command, Enki. Enki discovers the plot and goes to the supreme deity, Anu.

Anu confers with other gods (who remain nameless), and they bestow generalship on Marduk, with the promise of making him a god should his army succeed in thwarting Kingu's takeover attempt.

The gods are confident, and the Sumerian text describes a great party in the heavens with food and wine (I would love to be a fly on that wall). Marduk succeeds and wins the day. Both Kingu and Tiamat are destroyed. Marduk becomes a god and is known as the greatest god in ancient Babylon. No doubt there was some embellishment as Marduk was the supreme Babylonian deity.

"There was war in heaven..."
–Revelations 12:7

Notice that Anu (and the other gods) did not administer justice directly. They required the assistance of Marduk. Why? Were they not capable? After all, they were gods. Was their army not big enough? If their army was so small, why were they held in such high esteem? Being gods, didn't they have special powers?

I concluded, as I read the Sumerian text, that the reason must be that Anu, the supreme being, lived in another (spiritual) dimension and required the assistance of a physical entity from time to time.

It is interesting to note that this ancient text makes it very clear that it was the genes of the rebel Kingu that were used to create mankind. Are we therefore all related to Kingu? Is this another reason why Adam and Eve (the Adamu) were tossed out of the garden, so as not to pollute the genetically pure Anunnaki with the inferior genes from Kingu? Is this why we are rebellious by nature? Could this explain why the history of the world is a history of war?

This story suggests that perhaps war is not unique to this planet. Is it possible that war is a principal function of intelligent life in the universe? Perhaps good and evil are universal concepts and are at constant odds with one another. Hopefully, "good" will prevail in this struggle. At least until evil raises its ugly head again. Nonetheless, then comes the story of man's creation.

THE ENUMA ELISH
The Sumerian Story of Creation, Paraphrased

The Enuma Elish is the ancient Sumerian creation story. George Smith discovered it in 1876 while he was digging in the Royal Library at Nineveh (now northern Iraq). It is not without relevance that the ancient Sumerian creation story is told in seven tablets, paralleled by the Mosaic text in Genesis telling the six-day creation story. The first six tablets deal with creation, while the seventh describes the supreme Babylonian god Marduk bearing witness to his creation and then resting, just like the seventh day described in Genesis. The Sumerian story of Atrahasis "all wise" covers events much like the story of Noah and the Great Flood. These two writings date back to 1700 BC, and we suspect that the stories originated much earlier.

One important thing needs to be mentioned here. Some of the oldest Sumerian clay tablets, dating back six thousand years, describe our solar system, including the planets and their composition—gaseous, aquatic, etc.—as well as their colors and moons. How did the writers know this? They did not have telescopes, sophisticated NMR (nuclear magnetic resonance) instruments, or light-diffraction methods to help determine planetary contents. They did not send satellites to investigate other worlds.

Ancient interplanetary travelers must have told them. For further reading on this topic, Zecharia Sitchin's books *Genesis Revisited* and *The Twelfth Planet* are excellent sources.

The story goes on to describe an event that occurred 120 *sars* before the great deluge (a *sar* is defined in another Sumerian document, the King List, as 3,600 years). If one considers the Great Flood to be about 5,000 BC, the date in question is about 437,000 BC. At that time, Earth was visited by fifty Anunnaki ("those who from the heavens came"). They came from a home planet called Nibiru. Eventually, six hundred Anunnaki came to Earth, while three hundred of the Igigi ("those who observe and see")—also known as "the watchers"—remained in orbit about the planet.

IT STARTED SIMPLY AS A MINING EXPEDITION

In his books *Divine Encounters* and *Genesis Revisited*, Zecharia Sitchin describes how the Anunnaki home planet, Nibiru, was having a serious problem with atmospheric failure and required a lot of gold to protect it from the sun. This planet had an extremely elliptical orbit about our sun, lasting 3,600 years (one Sumerian sar). The reason we don't know of its existence is that most of the time it is so far away. (Although I admire Sitchin very much, I find this planet Nibiru's orbit theory a bit of a stretch.)

Nonetheless, this was a mining expedition. Anu was the supreme commander at Nibiru, and he sent his half-brother Enlil to Earth to supervise the gold mining, which took place primarily in eastern South Africa (known to the Sumerians as Abzu), where today there is still evidence of ancient gold mines. The Anunnaki on Earth sweated and toiled working the mines until, after 3,600 years, they could take it no more and rebelled.

The Rebellion of the Lesser Gods

No matter how technically advanced a people are, mining is hard and dangerous physical labor. Enlil was very frightened, because not only was he not getting the much-needed gold but he also had a serious and possibly violent rebellion on his hands. He summoned Enki, a more analytical and scientific person—one might say a bit more compassionate as well. Finally, they decided to go to the ultimate authority, Anu, on Nibiru. Anu sided with the rebellious Anunnaki workers and asked Enki to find a solution. Enki then had the bright idea of dedicating a biological worker to work the mines.

The obvious problem was there was no such biological entity available. The worker would have to be strong and bipedal. It would also need to be able to communicate, so as to take instructions and commands. No such animal existed. So a search was mounted to find the closest animal on the evolutionary scale and adapt it through gene manipulation.

ENKI AND HIS WIFE, NINMAH, GO TO WORK

Enki was bright, but his wife, Ninmah, was *very* bright and also skilled in this sort of biological enterprise. So she was also summoned to work on the project to create a "mixed one" or "Adamu." (Note the similarity to the biblical Adam.) The appropriate hominid was found, and through gene manipulation—inserting genes from one of the executed rebel leaders, Kingu—an intelligent worker was created to work the mines. Interestingly, the ancient text makes it clear that Ninmah volunteered to be the first birth goddess, and after nine months, it describes that a cesarean section type of surgery was required for the birth to take place. One could say that this was the first C-section in recorded history. And so a "mixed one" was formed to "bear the yoke" and work the mines. As planned, the worker was intelligent, strong, and communicative. It was made, as well, in the image of its creators. This seemed perfect, an ideal solution to the problem at hand.

The Anunnaki were both highly moral and ethical (or at least they tried to be). They knew the worker animal had to be intelligent but not too intelligent, or else it would be like enslaving their own kind. Also, so as not to interfere with the natural evolutionary development of the planet, the worker animal must not procreate and must have a natural mortal life-span consistent with other animals on the planet. This way, when the Anunnaki left, the new creature would just die out and not have to be destroyed.

There were seven male and seven female primitive workers, or "mixed ones," created with the help of Anunnaki birth goddesses. As stated in the first chapter, current DNA evidence suggests that HAR-1 (human accelerated regions) genes on chromosome twenty were inserted to form a bigger brain. A manipulated form of FOX P2 genes was then inserted on chromosome seven for language skills. These changes gave the animal a brain three times bigger than the most evolutionarily developed animal on the planet.

Then a small problem arose.

THEY MADE ONE ADAMU EVEN BETTER
Enuma Elish and the Bible

There is an ancient Sumerian text describing an event that scholars refer to as the Myth of Cattle and Grain. This text (although fragmented) describes early humans as eating and drinking without using their hands, much like any other animal. Also, they did not feel the need to wear clothes. In other words, they did not know they were naked.

With the consent of Anu, Enki and Enlil selected choice (more genetically advanced) workers and transported them from the mines in eastern Africa to Edin (yes, there is an ancient Sumerian city named Edin in southern Iraq) to work the garden and perform other, more domestic functions required there. They would enjoy a few perks, such as living longer, while serving as a sort of long-term pet that was able to cook and care for the place. (Notice that Eve is not yet in the picture. The reason, I suspect, is that at the time the Adamu were not able to procreate.)

> *And the Lord God planted a garden in Eden, in the east, and there He had put the man whom He had formed.*
>
> –Genesis 2:8

A problem soon became apparent. Enki and Ninmah had done too good a job. The Adamu came to realize they were naked, and they sought clothes. They began to eat with their hands (like the gods). Finally, it was discovered that they were also able to procreate, and then it really turned into a great mess.

The idea of these created ones having longevity like the Anunnaki was out of the question. They could multiply, and their aberrant and manufactured genes might take over or, worse yet, merge with the creators' DNA. The Anunnaki gene pool needed to be preserved and protected. So the Adamu had to be cast out, forcibly removed from any association, especially sexual, with the Anunnaki. They were now on their own.

> *But the Lord God called out to the man and said unto him, "Where are you?" He said, "I heard the sound of you in the garden and I was afraid because I was naked."*
>
> –Genesis 3:9

> *Then the Lord God said, "See the man has become like one of us, knowing good and evil and now he might reach out his hand and take of the tree of life."*
>
> –Genesis 3:22

So man was driven away from living with the Anunnaki. One chapter of mankind was closed, and a new one began. This new Adamu now had to survive on his own in the wild as any other animal.

There is an interesting statement in Genesis 3:21, in which God is said to have made "garments of skin" for Adam and Eve. Interestingly, he could have given them the fine garments from the Anunnaki closet, but instead he was more practical, showing them how to live on Earth and fend for themselves. After all, humans are not a pure product of evolution, as other earthly animals. We are "mixed ones," part evolutionary product and part manufactured, and thus lacking the survival-instinct genes needed to keep from becoming a genetic dead end.

I suspect the gods felt responsible in that regard and wanted to help. Perhaps God has been doing this all along, by sending heavenly messengers such as Viracocha in the Andes mountains of South America, Buddha in India, Confucius in China, Moses in Egypt, Osirus in Africa, Jesus in Palestine, Mohammad in Mecca, and so on, showing us how to live, how to survive, and how to relate to one another.

ATRAHASIS "ALL WISE" AND THE FLOOD STORY, PARAPHRASED

The world is full of ancient flood stories. The Greek and Romans had Deucalion, who was told to build a great ark because Zeus (Jupiter) was angered by man's wickedness and was going to send a great flood. The Chinese had a great water god, Gong, who created a great flood to punish the wicked. The Scandinavians, Africans, Northern Siberian tribes, Asian islanders, North American Indians, Persians, Indians, native Australians, and Mayans of Central America all have their flood stories, as do the Incas of South America, where the great creator god Veracocha saved mankind and the animals from a great flood. In that story, the Incan equivalent of Noah landed with his animals on Lake Titicaca, where life began anew.

Every corner of the world has a great flood story. The stories are all the same. Mankind became evil, and

God wanted to destroy or prune the genetic makeup through a great flood. Could it all just be psychological? Is it just something that is common in our human psychology? Or could these stories be an account of an actual event that took place after the great ice age ended, about fourteen thousand years ago, with the melting of the polar ice caps of the South Pole? Could these stories retell an actual single event, an event so big and terrifying that it left a permanent mark in human culture all over the world?

The clay cuneiform writing of the Atrahasis dates to about 1700 BC. It is very consistent with the biblical record, but it is more elaborate. Whereas the biblical account may be considered the *Reader's Digest* version, the Atrahasis is much more detailed.

Genesis chapter five gives us a list of ten great patriarchs before the flood. They include Adam, Seth, Enoch, Kenan, Mahalalel, Jared, Enoch, Methuselah, Lamech, and Noah. Interestingly, the ancient Sumerians also gave us a list of ten great kings prior to the flood. These tablets date to 3200 BC.

Nonetheless, paraphrasing again from the ancient text, humans were thriving on Earth and multiplying in geometric progression. Enlil gets tired of the "noise"—no doubt a reference to the constant reports of the violence and cruelty perpetrated among humans. So after twelve hundred years, he attempts to wipe them out through a plague. This fails.

After another twelve hundred years, he attempts the same thing by rain. After another twelve hundred years, he attempts again to destroy mankind, this time by drought. Each time he is foiled in part by Enki, who gives food or water or medical help to the humans.

Finally Enlil says enough is enough and seeks to destroy all of mankind through a massive flood. To be sure of success and to be sure that his plans are not foiled again, he makes everyone, including Enki, take an oath of secrecy.

So this is how we get to the story of the Great Flood. But is it just a story, or is there science behind it?

A POSSIBLE SCIENTIFIC EXPLANATION FOR THE GREAT FLOOD STORY

As described in the journal *Science* (January 15, 1993) and later in *Scientific American* (March 1993), a huge chunk of ice on the polar ice cap of Antarctica could have been dislodged after the unsettling climatic changes of the last glacial epoch. This could have caused a supercolossal tidal wave estimated in excess of fifteen hundred feet tall. A terrifying wave of water this size would certainly destroy and kill anything in its path.

Thus a watery holocaust of biblical proportions was potentially set loose, causing a tidal wave of destruction up the Persian Gulf and obliterating humans and livestock. No earthlings could have seen it coming, and not even Enki was able to tell them.

ENKI SAVES MANKIND

Enki was very troubled. He wanted to save mankind, but he'd taken an oath not to tell. So he did something very sneaky.

He found Atrahasis (Noah) and stood just on the other side of a reed wall, talking loudly to an imaginary person so Atrahasis would overhear. This way, he did not break his oath. You can imagine him saying, "Boy, if I were Atrahasis, I would build a big ark and save myself and my family. I would also save as many of the animals as I could, too, because a big wall of water is coming. This is how I would build the ark..." Astrahasis overheard, and you know the rest of the story.

The flood was a more terrible event than even Enlil had bargained for. When he discovered he had been foiled again by Enki, he was angry at first, but then he was pleased that Atrahasis and his family had survived. Enlil promised that he would never again interfere like this with the affairs of mankind. Why was he so pleased?

DNA AND
THE GREAT FLOOD

Is this the way it happened, or is there more to the story? There are some statements in Genesis regarding Noah that always stuck in my head. Genesis 6:9 states that "Noah was blameless in his generation" (other versions say "pure"). This is an obvious reference to the fact that he was not only a good man but also of good genetic stock. Remember, the beginning of Genesis chapter six refers to the sons of God having sex with the daughters of man. This obviously resulted in a race of cruel and violent demigods and a complete genetic screwup.

I remember years ago asking a priest what "sons of God having sex with the daughters of man" meant, and he just gave me the party line: "Some things are not meant to be understood." Well, as you can surmise, I never accepted that point of view. It is there and was written down thousands of years ago for a purpose.

The purpose is that it be read and, hopefully, someday understood. As we will see, the study of other extra-canonical scriptures begins to clarify this mystery.

At any rate, Noah was of good and pure genetic stock. Considering this fact and the multiple efforts at population control, one can conclude that these plagues and droughts, etc., were attempts at genetic pruning, and the Great Flood was the ultimate attempt. All the genetic crap was to go, and the seed of Noah, which goes back uninterrupted to Adam, would be saved so that Earth would be replenished with good genetic stock. This is the reason Enlil was so pleased that Atrahasis (Noah) survived.

However, we still have evil. Is it possible that the current evil is the result of aberrant genes and insufficient pruning? Is more pruning needed? (God, I hope not...) Or is it just the proper balance of yin and yang, and evil just *seems* overrepresented?

SEX AND
THE GODS

In Genesis chapter six, the story is told that there were "giants" in the days before the flood. The Hebrew word for *giants* is *Nephilim* (meaning not necessarily physical giants but "those from the heavens came"). It was the Nephilim who looked upon the daughters of man, saw that they were fair (beautiful or even sexy), took for themselves wives of their choosing, and had offspring. The Bible goes on to describe the offspring as the heroes of old—men of renown, great warriors. It also goes on to describe a corrupt world filled with evil.

Interestingly, Genesis 6:4 says that the "Nephilim were upon the earth in those days...and after too." This is to say that these beings, these *ancient aliens*, were on Earth in those days and—oh, by the way—*afterward, too.* Like maybe they never left.

So beings from the heavens came to Earth, mated with human women, had children from those unions, and, at least for a time, stayed on the planet. It is there in the Bible for all of us to read.

THE BOOK OF
ENOCH
Another Story of Sex and Violence

Enoch was the seventh patriarch before the Great Flood, the great-grandfather of Noah, and the grandfather of Lamech. His book is written in the first person, so the writer claims to be Enoch himself. Although some fragments were preserved in AD 792, the book was lost until James Bruce in Abyssinia obtained three copies of the Ethiopian version in 1773. The copies were written in the Ge'ez language and dated to 300 BC.

So this book could not have been entered in the biblical canon, as it was lost to antiquity. It is, however, part of the Ethiopian Orthodox Church canon, and is also known as the Book of the Watchers. It was highly revered among early religious leaders (both Christian and Jewish). Clement of Alexander (an early Christian writer) quoted from Enoch as if it were scripture. The kabbalah teaching book *Zohar* refers to the Book of Enoch as carefully preserved from generation to generation, starting with Enoch himself. Origen (the most prolific early Christian writer) said in AD 259 that the Book of Enoch had the same authority as the Psalms. It should also be noted that Greek and Aramaic fragments of this book were found in the Dead Sea Scrolls. These fragments matched the Ethiopian text rather well, indicating its authenticity.

There are primarily three interesting things described in the Book of Enoch.

1. Chapter VII, Section II
   The book amplifies the sixth chapter of Genesis and describes in more detail the sexual transgressions of the fallen angels. It names the perpetrators and describes their occupations and skills. It tells how they did what they did and where on Earth they entered. It informs us that

their leader's name was Samyaza and that he took with him two hundred confederate angels and descended on Mount Hermon during the time of the sixth patriarch, Jared, with the intention of breaking God's law and having sex with the female humans. The product of these sexual affairs was the demigods, and what a violent mess they were. The souls of the bloody victims cried out unto God. So God became sad that he created mankind and planned their destruction in a great flood.

2. Chapter CV

There is also an interesting story about Lamech, the father of Noah. As the story goes, when Noah was born, his hair was "white as snow" and his skin was "red as a rose." His hair was "white like wool and long." "When he was taken from the hand of the midwife, he opened his mouth and blessed the Lord of Heaven." Lamech believed that the child was not his, because he resembled the "angels of heaven." So he went to his father, Methuselah, who in turn went to Noah's great-grandfather, Enoch, who by that time was living among the Watchers in the heavens. Enoch assured Lamech that the son was indeed his and that his name was to be Noah, for he would survive the Great Flood that was to befall the Earth soon. His seed would replenish the Earth.

Was Noah the product of artificial insemination to fulfill a genetic goal? If so, this was more than just pruning out the bad genes. This indicates an active process of introducing newer and more acceptable genes so the future human generations would be even more like the Anunnaki. To what purpose and to what end? Is mankind simply part of some intergalactic genetic experiment? Is it possible that every so often the gods come and do a genetic tweak here or a tweak there and then wait a few thousand years to witness and document the results?

3. Chapters LXIX and LXX

   The book also describes three primary god figures: the Ancient of Days, Son of Man, and the Lord of Souls. One may consider this a trinity, for that is exactly what it is. The book goes on to describe the importance of the soul, be it wicked or otherwise. It also describes the souls of the fallen angels and their offspring as being tied to Earth and as evil spirits. Clearly, the Lord of Souls holds a very special and elevated level of authority.

(It should be noted that a similar story is given in the Book of Noah, which some scholars consider to be even older than the Book of Enoch. It should also be noted that portions of this lost Book of Noah were found among the ancient manuscripts known as the Dead Sea Scrolls.)

THE BOOK OF THOMAS
Clarification of the Trinity and How
It Relates to Enlil, Enki, and Anu

Nothing was known of this book prior to 1945. Then, Egyptian peasants digging for fertilizer discovered the thirteen leather-bound manuscripts stuffed in a jar. It has been estimated that these manuscripts were buried around the fourth century AD. Because they were found near the village of Nag Hammadi, they became known as the Nag Hammadi Texts or Library. The writing was in an ancient form of Egyptian known as *Coptic*, although it is believed that the original was in Greek.

The most interesting manuscript in this jar was the Book of Thomas. This writing contained 114 statements purported to be directly from the mouth of Jesus and transcribed by a certain Didymus Judas Thomas. (Ancient legend also purports this person to be the twin brother of Jesus.)

The writing describes no miracles. There is no talk of the painful and humiliating death or the resurrection of Jesus, only his sayings, which are described as "secret" phrases. Nonetheless, it may be possible that the Book of Thomas is the most accurate text we have of what came out of Jesus's mouth.

There are many interesting verses, including verse twenty-nine, where he explains that this Earth is actually a birthplace for souls. "The spirit came into being because of the body." However, the most revealing text, in my mind, is found in verse forty-four, where Jesus states,

"Whoever blasphemes against me or my father will be forgiven, but whosoever blasphemes against the Holy Spirit will not be forgiven on Earth or in heaven." This clearly states that the Holy Spirit of the New Testament—known as the Lord of Souls in the Book of Enoch—is the top figure. This explains why monotheism in the Jewish and Moslem tradition is consistent with scripture.

It also explains the controversial use of the plural *Elohim* (gods) in the Genesis account of creation: "Let *us* create man in *our* image" (Genesis 1:26). There are two physical persons that may be described as God (Enki and Enlil), but the big guy that is constantly in the spiritual realm is the ultimate GOD (Anu). This makes perfect sense because of the permanent and deciding nature of the spiritual realm. All physical beings will someday have to go there. It is there that your future fate is determined. It is there, in this spiritual realm, that the disposition of the soul is made. The ultimate judge. This explains that, yes, there is only one god, but at the same time there are others as well. Perhaps there are multiple physical gods—which some might describe as ancient aliens—but only one in the spiritual realm, and this one is the ultimate authority.

Furthermore, all ancient texts in cuneiform script state that Anu resides in *heaven*. Zecharia Sitchin states, however, that Anu resides on the (hitherto undiscovered) planet Nibiru, whose extremely elliptical orbit around the sun takes 3,600 years. Any astrophysicists will tell you that such an extremely elliptical orbit, for a planet, is impossible in the physical universe. The first orbit will be its last, because the planet will clearly achieve escape velocity, flinging itself out of the solar system, never to return. Thus, if Anu truly resides on Nibiru, then he resides not in the physical universe but in the *spiritual* realm.

Furthermore, the word Nibiru in cuneiform is symbolized by a cross and occasionally by a winged disc. The word translates to *a crossing,* such as crossing a river. Ancient Greeks and Romans would put a coin in the mouths of their dead loved ones to pay the ferryman who aided the soul in *crossing* the river to *transcend* to the other side. Anu, then, is the supreme god that lives on the other side of the "river," and in order to get to him, one must *cross* that river to *transcend* to the spiritual side of existence. Anu in this way is akin to the Holy Spirit as described in biblical scripture and the Great Spirit as described by the American Indians.

IN SUMMARY

Putting all this evidence together, ancient texts suggest that humans were created for the gods to serve as miners and domestic help in the homes of the Anunnaki. To create us, the Anunnaki manipulated the genes of the highest evolved bipeds (possibly *Australopithecus*, a bipedial creature with a brain at one-third the capacity of humans) on Earth at that time (400,000 to 250,000 years ago).

This would explain why the Egyptian hieroglyphic symbol for the gods was the pickax. In addition, the

Egyptian word for *gods* is *Neteru*, which also means *guardians*. It is also not without relevance that ancient Sumer's common local name among the ancient local peoples was *Shumer*, meaning "land of the guardians." Thus the root of ancient Egyptian mythology emanates from Sumer and Mosaic teaching, making the connection (another golden thread) that comes all the way to this day. Sumerian → Egyptian → Judaism → Christianity. It is all tied together, and these old myths deserve some thought. These writers of long ago believed that they were not just creating entertainment but recording real events relevant to human history. Our ancient ancestors were trying to tell us something about our past that can help answer the three major questions of life: *Where did we come from? Why are we here? Where are we going?*

The humans were formed from DNA from the rebel Kingu, no friend of Anu, the supreme deity. Thus an intercessor was needed for their preservation and who would be better suited than Enki and his wife, Ninmah. After all, they were responsible for these *Homo sapiens*, and Ninmah was their "mother."

But first we need to know a little bit more about ourselves. We need to understand the origin of human consciousness. If we can do that, then we should be able to understand the world of the Anunnaki (the world of the gods) a little better.

CHAPTER THREE

HUMAN CONSCIOUSNESS
A Scientific Case for the Soul

The true lover of knowledge is always striving after being... he will not rest at those multitudinous phenomena whose existence is in appearance only.

–Plato

ME
Who is the man
Behind these eyes?
Who sheds the tears?
Who laughs and cries?

Is not the brain
The mind?
But who's what
And what's mine?

HUMAN CONSCIOUSNESS

Am I simply neurocircuits
Of an electrical scheme?
Or is there a life
Deep inside, yet unseen?

Nay, says the skeptic
For you are a machine
And the concept of a soul
Is but a dream.

For you are you
And the you is your brain
That laughs, cries
And feels the pain.

But I say at risk
Of sounding dumb,
What part of the brain
Does the question come from?

For when the brain is divided
And all parts identified,
Where's the soul or self
That's gone when I die?

I am more than this.
More than nerves and flesh.
More than face, eyes, and voice
And all the rest.

For when all is gone
And no flesh to see,
There will be something
And that is me.

–David D. Weisher, MD, AD 2000

The study that we are embarking on in this treatise can never be complete without discussing the nature of human consciousness. It is at the heart of all philosophies and, of course, religions. Yet, is there any evidence that consciousness may indeed be spiritual, as the ancient philosophers and religious leaders proposed for thousands of years?

I, for one, did not believe so. It was my secret persuasion for many years that anyone who believed in a soul or spiritual origin to human consciousness was intellectually inferior. I would neither discuss it nor be interested in talking with anyone who held such a position, for clearly such people were inferior, I thought, and would likely bore me to death.

Nonetheless, the nature of human consciousness has always fascinated me. As a child, I remember flashing a light into my pupil and witnessing it constrict in response. I thought, "What an amazing machine we humans are." There was no room in my mind for any spiritual side, only for the amazing human cerebral machine.

One of my early careers was as an aerospace computer engineer. I was privileged to help develop the computers for the F-14 Tomcat (a primary air superiority aircraft for the navy) and B-1 Bomber (a primary bomber aircraft for the air force). Most of my attention was on the F-14. My specialties were the computerization of wing sweep and the transition from subsonic to supersonic flight. If you ever wondered why the Iranian F-14 planes never flew in any hostile action, it was because I had their computers (CADC and AICS) in my lab in North Torrance, California, in the late 1970s. (The shah bought about seventy F-14s.) I suppose I'm setting myself up as some sort of target by revealing this, but I'm old and becoming less relevant as the years go by. Nonetheless, I believe I did the right thing by keeping the computers in California and thus the Iranian jets on the ground.

The F-14 started with only three primary computers, but it was designed to be modular and thus became very flexible as computer capabilities were constantly added. As the computer technology advanced, I began to be more fascinated with the nature of human consciousness. I saw the machines as having more "thinking" capabilities, which caused me to wonder if someday a machine might mimic or achieve a form of consciousness. I could not get this idea out of my mind.

Then one day someone told me about this new amazing machine that could take a picture of a living human brain. It was called a CAT scan. So, being young, single, and flush with cash (it was just writing checks and adding more zeros, not rocket science), I went to medical school. Oh yes, I considered psychology, psychiatry, and neuroscience, but I believed that only in the medical field of neurology would I receive the objective and scientific perspective of human consciousness.

So off to school I went. I completed my neurology training at Georgetown University in Washington, DC, in 1988. I then went on to become board certified in neurology, sleep medicine, and undersea and hyperbaric medicine. I practiced in the Washington, DC, area for many years, working with six hospitals, six intensive care units, a trauma center, and a very busy private office. With a practice that busy, some profoundly interesting cases regarding near-death experiences (NDE) are sure to come by, and many did. So as not to bore you with endless near-death stories, I chose the three that are most compelling and that convinced me to start thinking outside the dogma box.

HEAVEN'S WITNESS

It was around 1993, during my practice in the Washington area (Greenbelt, Maryland, to be exact), that I had what one might describe as a life-altering event or epiphany. I was at Doctors Community Hospital in Lanham, Maryland (just down the street from my own office), when I was requested to go to the emergency room. There on a stretcher in a small cubicle lay a male patient (in this story I will call him Patient 1). I was told that he had simply been walking down the street when he had a seizure, which put him in a state of cardiopulmonary arrest (in which the heart stops beating and the patient stops breathing). CPR was being administered. In other words, he was dead, and we were attempting to revive him in the emergency room.

We got his heart beating, but he was still unconscious. So we transferred him to the intensive care unit

(ICU). Over the next few days, he began to regain consciousness. He recovered fully a week or so later and was discharged.

During his hospital stay, I ordered a CAT scan of his brain to find out why he'd had a seizure in the first place. (I almost never do CAT scans with IV contrast, because it risks side effects including renal impairment and even allergic reactions resulting in death.) The CAT scan of the brain was normal, so on discharge, I ordered an MRI of the brain with contrast (MRI contrast studies are much safer) and had the patient schedule an appointment in my office for a follow-up visit.

When he came into my office, he sat down in front of me, with his wife sitting in the chair on his left side. He said, "Dr. Weisher, I was dead." I replied that I knew he was dead, because I was there, in the emergency room: no breathing and no heartbeat. He then said, "I was out of my body and saw everything, including you in the room, and I went to the other side."

He then went on to describe in vivid and accurate detail the events that took place in the little cubicle in the ER when he was unconscious and in cardiopulmonary arrest. Yet at no time while he was in that cubicle was he ever "physically" conscious. He described a light and how he'd felt a strong desire to go to it. He said that when he did, it was like traveling faster than the speed of light. He then found himself in a beautiful landscape. He described the place as a beautiful green field with a river coming down the middle. Off to the left was a very distant mountain. Surprisingly, he said that he could focus his vision on an individual branch of a tree on that mountain and his hearing on a particular bird singing. Across the river there was a big tree, and some people were gathered there underneath the tree; they seemed to be waiting for him.

At this point I interrupted him, saying this was a hallucination brought about by an encephalopathic brain and resulting in the random firing of neurons that create a visual and auditory illusion of reality. I must admit that I felt rather smug and confident about this. After all, I was the doctor, and he was only a patient. What could I possible learn from him? Then he said something I will never forget.

He said, "Dr. Weisher, I'm older than you, and therefore I've had more dreams than you. This was no dream. My vision was better. My hearing was better. My thinking was very clear. I asked questions, and I received answers. It was much more real than you in front of me today, and that is what I can't get out of my mind. So don't give me this bullshit about Donald Duck or spiders on the wall."

So I shut up and listened to what he had to say. He went on to describe a visit with his dead uncle and father. Then a man tapped him on the shoulder and said he was Jesus. The patient turned to him and said, matter-of-factly, "Well, I don't believe in you." Jesus just laughed and said, "Well, I was a historical figure and just wanted to tell you that you lived a very exemplary life and you should be proud."

Now the man's dead brother was there also and wanted to show him his new sailboat. While on Earth, the brother had been addicted to sailing and talked of nothing else. The sailboat was at dockside, and the man described it as approximately 120 feet long with lifelines encrusted with all kinds of jewels, thousands of diamonds, rubies, emeralds, etc. It was like no sailboat he'd ever seen.

He then found himself sitting in the cockpit of this sailboat. His brother was at the helm with his hand on the wheel. He could see clearly as never before the stars of the universe. Patient 1 then told his brother that this place was so incredibly wonderful, he just never wanted to leave. His brother then told him that he would have to leave and tell this story to others, so they might know the reality of existence. He also said, "But don't worry. You are only going to be on Earth for a short period of time, for you are terminally ill and will soon come back." Patient 1 thought, "What kind of wonderful news is this? Oh boy, I'm terminally ill and going to die!"

Now this story is *absolutely true*, and while he told me this, I was putting up the brain MRI film in the viewer for inspection. On the MRI for all to see was a glioblastoma multiforme grade IV malignant brain tumor. He knew he was terminally ill before coming into my office, before anyone on Earth knew it, because they in the hereafter had told him. Every time I think of this event, the hairs on my neck become stiff.

At the time of his diagnosis, I was on the board of trustees for the Prince George Hospice Society. Knowing this, he requested permission to visit the very sick and the dying, to tell them that they were going to a wonderful place and would not be forgotten. His wife almost fell off her seat. She said, "I no longer know this man." She then informed me that Patient 1 was a businessman first and foremost, always looking for the great deal. (His business was real estate, and he was, by all accounts, extremely fair and honest.) He had no religion or any spiritual background whatsoever and no interest in spiritual topics.

Patient 1 informed me that he wanted no heroic measures in his care for his brain tumor. "I know where I'm going." So no surgery was performed, and only radiation and steroid treatments were given. He died, six months later, the way one should: at home and with family.

His son told me the story of his death. They gave him a puppy to keep him company. The dog would always bark at strangers. Just seconds before Patient 1's death, he said that he saw his father and brothers coming to meet him. At that time, the puppy looked in the same direction and started barking at someone in the corner of the room. Needless to say, no one was there. "At least no one anyone could see," his son said.

I wrote this book in part to tell his story. I have no reason to lie or bend the truth. I have no profit in fiction. I just follow the data wherever it leads me and hope in some way you might benefit.

This patient clearly educated me. I did not change my philosophy then and there, but he certainly gave me food for thought. My curiosity took over, and I began my own research on the topic of consciousness. I concluded that he was right after all.

We are spirits in a temporary physical form. It is for this reason I wrote my second book, *Mysteries of Consciousness*. Although I'm not religious, when someone in the hereafter describing himself as Jesus Christ says to take his message to Earth, to profit from that message is simply not righteous. I have not been a good steward of this information, because marketing is my weakness, and so I must apologize.

Now, you may ask, why is the nature of consciousness so important in our discussion of this golden thread? It is not only important but also critical. If all there is to the nature of "self" is neuronal function, then we can stop right here and go no further. But if consciousness is "extraneuronal," then existence does not expire with death. If existence does not expire with death, then there are other realms of existence and, therefore, other levels to this hierarchy, beings of influence in other dimensions that have input in our physical existence. So one cannot ignore the study of consciousness when contemplating the great philosophical questions of life.

As I've said, I've had the opportunity to bear witness to many patients with near-death experiences and am now personally convinced that most of these experiences are real and present a doorway to another existence. Einstein once said that "space and time are facts that we live by but not fundamental principles of the universe." Our existence is clouded by limited vision, limited hearing, and limited concentration. All of my near-death patients described clearer vision and thought, a greater sense of reality, a feeling of increased liberty. They often say things like, "Getting out of my old body was like taking off an old coat."

THE CASE OF TWO
DEPARTING SOULS

I recall an interesting case that took place at the Schneider Regional Medical Center in St. Thomas. This took place around 2005, about two years after I came to the Virgin Islands. I recall getting a frantic call from the ICU nurse at about three in the morning. She said that my patient was screaming nonsense and the staff could not calm him down. He was suffering from hepatorenal syndrome, a rapidly deteriorating and ultimately fatal condition involving the failure of both the liver and kidneys. I could hear him yelling in the background, "Turn her loose! Turn her loose! Let her go!" I ordered more sedation and came into the ICU early that same morning.

When I arrived at his room (I remember the room was twenty-four), I could still hear him yelling, "Turn her loose! Turn her loose!" I sat down next to him to try to glean the reason for this excitement. I will never forget what he did next.

He looked straight at me and said, with convincing clarity, pointing to the next room, "There's a woman next door, and she had a bad heart attack, and she has come out of her body to tell me that she is between this world and the next and wants to be turned loose so she can go to the next world. *So turn her loose!*"

How did he know there was a patient next door in room twenty-five? How did he know she was a woman? How did he know she'd had a heart attack? And most importantly, how did he know she was near death? The woman in room twenty-five had suffered cardiopulmonary arrest and was very near death, in a coma. In fact, her EEG (electroencephalogram, or brain wave test) showed a burst suppression pattern, indicating impending total cerebral death. It was convincingly clear that she did the only thing she could to get the message to us and that was to inform someone else who was near death (a better term would be *near transition*) and therefore in the same plane of consciousness. That is exactly what she did.

A few days later, both went to the hereafter.

THE CASE OF THE DOG
Dogs Have Spirits, Too

The concept of a soul may not be limited to only humans. Man's best friend may also be included in this club.

When I was practicing in Maryland, I had two large and beautiful Chow Chow dogs. They didn't cost anything because they were saved from the "execution dock" at the local Humane Society. Their time had run out, so to speak. The big and beautiful black one was named Shogun. Ginger was blond and much older but still beautiful.

One day, Ginger was obviously not feeling well, so she was taken to the veterinarian's office about ten miles away in Annapolis. She had a tender mass on her front leg. The doctor called and stated that he felt it might be serious, possibly a malignancy. Then he called about an hour later and said it was only an abscess, which he drained. He informed me that she could be picked up sometime the next day.

That night at about two thirty, Shogun released the most awful and terrifying sound I had ever heard from her. It was bloodcurdling. These dogs were normally quiet, unless something very unusual had occurred. My initial thought was that someone broke into the house and stabbed the dog. However, it soon became obvious that this was not the case. He just howled and howled until, exhausted, he finally went back to sleep.

That morning I got another call from the vet. Ginger had died. I asked when she'd passed away. He said about two in the morning. I said, "That's incredible, because Shogun was howling very loudly at that same time." He then informed me, rather perfunctorily, that this behavior is not infrequent; these animals have a sixth sense and can detect the departing soul. I thought to myself, "These are not dumb animals after all. Is it possible that man's best friend also has a soul?"

If so, what are the implications? Do these animals have souls because of their association with humans? Because of their relatively advanced stage of cerebral evolution? Or do they just have souls because they're alive?

There may be so much more to existence than we can ever understand. Maybe what Jesus said in the book of St. Thomas was correct: that the soul comes into being because of the body (perhaps including dogs). Those souls are continually being created as the population expands.

FOR THE MEDICAL RECORD
NDE in Medical Literature

Medical science is now getting into the field of near death. Not too many years ago, any discussion of the near-death phenomenon was almost a career killer, because many of our scientific leaders considered this a discussion in religion, and thus it was not acceptable for publication. Some of my dearest friends in the neuroscientific community had to suffer the wrath from their "inquisitions." I guess, to paraphrase a New Testament phrase, the ignorant ye have with you always.

The scientific argument will be discussed later on. Let us now focus on one particular study, published in the *Lancet* journal (2001) and written by Dr. Pim van Lommel, et. al., from the Division of Cardiology, Hospital Rijnstate Arnhem, the Netherlands. This was an interesting study of 344 consecutive patients successfully resuscitated after cardiac arrest in ten different Dutch hospitals. They were evaluated for NDE and for possible psychological changes after such an event. One of these patients had a most interesting experience.

A forty-four-year-old male was found comatose in a meadow. He had suffered a myocardial infarction (heart attack) resulting in cardiopulmonary arrest. He was brought to the hospital emergency room. While there, in a small cubicle, he was given nonintubated ventilation and cardiac massage. The decision was then made to intubate him (put a long tube down his trachea, connected to a ventilator for respiration control), but it was discovered that he had dentures; these had to be removed prior to intubation. The dentures were removed, and the patient was successfully intubated and sent to the ICU, still comatose, although his heart was now functioning.

A few days later, the patient regained consciousness and was extubated (taken off the ventilator). Now it was time to give him his dentures. However, they were lost. Despite an extensive search, they were nowhere to be found. The patient was transferred to the cardiac step-down unit and, of course, had to eat soft food. One day, he was visited by the nurse who had taken care of him while he was comatose in the emergency room. The patient immediately recognized her and explained that he had been out of his body while receiving cardiac massage and artificial ventilation, and he saw her put his dentures in the "crash cart" in a "drawer with all these bottles." The dentures were put "under a sliding drawer" within that drawer.

The entire time this patient was in the ER, he was not only comatose but also clinically dead (no cerebral function). Yet, apparently he was able to maintain a form of consciousness outside the normal neuron-to-neuron epiphenomenon of cerebral function. One might say, though he was dead, he still lived.

The dentures were found in the crash cart as the patient described. Clearly, this is compelling evidence of human consciousness originating outside the known neuronal sphere, evidence suggesting that human consciousness may not be neuronal after all but extraneuronal. It may, in fact, have very little to do with the brain. While this man's brain was utterly inert, he saw (and could later recall) the room, the nurse, and her movements.

I am unequivocally and professionally convinced that this big mushroom we call the brain is not the originator of human consciousness but an *integrator* of our spiritual consciousness. It allows our true, bodiless consciousness to interface with this four-dimensional world we currently live in.

Yet there is another aspect of medical science that now confronts physicians with this enigma of thought origin. There is a major medical/surgical procedure in which the brain is actually brought to a state of cerebral death and returned to life again, a state in which the brain clearly and unequivocally shuts down. And yet patients undergoing this procedure report a vivid form of consciousness.

THE CARDIAC STANDSTILL OPERATION
Everyday Evidence for a Soul

A cerebral aneurysm is a bubble on one of the arteries of the brain. Although relatively small ones tend to be benign, larger ones (greater than ten millimeters in diameter) and giants (greater than twenty-five millimeters in diameter) can be very dangerous. Giant cerebral aneurysms often result in catastrophic intracranial bleeding, without warning, causing death or severe, permanent neurological sequeli. Fifty percent of people suffering these aneurysms die. For this reason, patients with giant cerebral aneurysms are desirous of some sort of correction, rather than living with that sword of Damocles over their heads, so to speak.

Radical situations often require radical solutions. Giant cerebral aneurysms cannot be simply clipped like smaller ones, and conventional craniotomy would be so radical and so long that survival would be doubtful. The slightest cut to the artery bubble would release so much blood that the surgeon would be unable to see, and the patient would be at risk of bleeding to death or suffering a severe and permanent neurological sequeli.

The alternative is cardiac standstill surgery. The procedure is fairly straightforward. The patient is sedated and then cooled to less than 58 degrees Fahrenheit. When the heart stops beating, the blood is drained from the body. This enables the surgeon to operate with a "clean field" of operation. The surgeon now has thirty to sixty minutes (though ideally the procedure lasts less than twenty) to cut and sew the aneurysm with no problem visualizing the surgical site or risk of blood loss.

During cardiac standstill surgery, there is neither pulse nor respiration. The EEG is completely flat during this procedure, as the brain is receiving no blood, oxygen, or glucose for fuel. Because the brain has no significant storage for its fuel, it quickly shuts down, putting the patient in a state of, essentially, suspended animation. That the body is cooled down to less than 60 degrees Fahrenheit allows survival from such an extreme state of cerebral shutdown. (This survival "window" has been known since the 1960s, through studied cases of hypothermia.)

A person in cardiac standstill surgery actually undergoes a sort of controlled state of near death. From a neurological perspective, the patient has achieved a state of true and total death, through total cerebral shutdown. Anyone who disbelieves NDE because of the often-quoted adage, "Anyone who came back was not truly dead," does not understand the neurophysiological implications of this procedure. The brain is clearly and unequivocally shut down, and yet patients are describing an incredible form of consciousness. This cannot be ignored or swept away. This is extremely relevant.

Michael Sabom, MD, is the cardiologist who performs cardiac standstill operations at the Veterans Medical Center in Atlanta, Georgia. He has borne witness to many near-death events and has written several books on the subject. To him, his association with this procedure has been life altering. There are well over two hundred incredible NDEs in this relatively rare and infrequent procedure, which is very much consistent with Dr. Melvin Morris's finding that to have an NDE one must be truly near death, not just sick. (I can confirm this from personal experience as a neurologist.)

Pam Reynolds was one patient who underwent such a procedure and had a near-death experience. Her surgery was performed at the Burrows Neurological Institute in Phoenix, Arizona, one of five or six medical centers capable of doing cardiac standstill surgery in the United States. She accurately described procedural details and events of her surgery as viewed from above, outside her body. She described perfectly the type of instruments used and the sounds they produced as well as the hats the medical staff wore, despite the fact that her brain was completely flat-lined and she should have had no consciousness whatsoever. She was able to learn aspects of the procedure completely unknown to her before, as she had never seen these surgical instruments or been familiar with the procedure prior to surgery. She accurately described who said what and who wore what while she was clinically dead.

Pam then saw a light at the end of a tunnel and went through the light at an amazing speed. At the end of this tunnel, she visited dead relatives. She described the experience as "very real" and more visceral than normal existence. Like many other NDErs, this was a life-altering event, and she was spiritually changed forever.

No neurologist can deny the great neurological paradox that during a state of complete cerebral inactivity, a heightened and more real form of consciousness—apart and independent from the brain—is achieved. It should be noted that the term *near*-death experience (NDE) is actually a misnomer. Studies and personal experience make it very clear that this is a *death* experience, and *near* has nothing to do with it. Therefore, it should be termed death experience (DE). I believe unequivocally that these *near-death* experiences are a window into another form of consciousness that is higher than the one we currently experience, a consciousness that is free from the limiting physical encumbrances that we face today. No, my friend, neurons are not the origin of human consciousness but rather a limiting facilitator allowing a temporary experience with this primitive form of existence. (A more complete dissertation on the subject of consciousness may be found in my 2005 book, *Mysteries of Consciousness*.)

And if we, with our relatively meager level of technical/scientific development, can deduce that the origin of human consciousness is extraneuronal (spiritual/soul based), then our ancient alien visitors must also know this, and it must be an integral part of their world. For the latter part of this chapter, I elected to discuss some of the accepted scientific theories regarding human consciousness and debunk them.

SCIENTIFIC EVIDENCE FOR EXTRANEURONAL CONSCIOUSNESS
The Academic Argument for Human Consciousness

To pursue our argument for an extraneuronal source of human consciousness, we will have to delve into various scientific theories and postulates. I apologize if I become too scientific for you, but this is necessary to satisfy some of my more scientific critics. If I don't get specific, then my academic colleagues will counter with the old and somewhat boring theorems, hypotheses, and other dogma that I've heard a million times and know to be unworkable and, in some cases, even ridiculous. Anyway, if this scientific jargon is not your cup of tea, you may skip this section and go to the next chapter. I suggest, though, that you give this section a try at some point, as it is relevant.

***

If there is a neuroscientific orthodoxy regarding human consciousness, it stems from the turn of the last century. William James, often considered the father of American psychology, famously said, "Human consciousness is a process and not a thing." He postulated that human consciousness is the result of the epiphenomenon of synaptic firing neurons—that is, simply, the sequence of cerebral electrical activity and not the cerebrum itself. In other words, if one is to do an exhaustive study of human consciousness, one need go no further than the human brain.

Dr. James may not have been the first with this idea. Rene Descartes (the French philosopher, 1596–1650) famously said, "I think, therefore I am." Thus the *process* and not the *thing* is what our scientific leaders focused on when dealing with the subject of consciousness.

When giving talks on this subject, I often refer to this concept as the theory of *singularity*: the brain, and only the brain, is what creates human thought. If you have ever taken a college course in psychology, you surely remember B.F. Skinner and the philosophy of behaviorism. This is the same as singularity, for behaviorism is primarily interested in a sequence of physical actions. If I like Mars bars, and you give me one every time I press the red button, then I'm going to press the red button often.

The ancient Greek philosophers had the opposite view, believing that the true origin of consciousness was a thing, and in fact this *thing* was separate from the brain. (They perfectly understood that humans thought with their brains and not with their livers, as the Babylonians thought, nor with the heart, as the Hebrews believed.) Ancient Greek thinkers (including Pythagoras, Diogenes, Aristotle, Plato, and Socrates) believed that there were two sources of human thought: a temporally (mortal) physical brain and a permanent, immortal, nonphysical soul that goes on after death. It is for this reason that this line of thought is referred to as *dualism*.

The singularists (if I may coin that term) believe that the illusion of a soul arises as a result of the sophistication of the brain. As the brain becomes more complex (that is to say, as the brain evolves), individual and independent thought will arise. In time, the individual will question the origin of that thought and, hence, give birth to a concept or theory—some would say *illusion*—of the soul.

THE TURING TEST

One is reminded of the famous "Turing test" of 1950. Alan Turing was a brilliant British mathematician who, during World War II, was instrumental in the development of the computer. Computers were very primitive then, but people began to wonder whether it was possible that these new "brains" would ever have a consciousness. So Turing devised a simple scenario.

He thought that if you placed a person in one room with a keyboard terminal and a computer in the other room and fed them each questions, one might just mimic the other. In other words, if the computer was sophisticated enough, it would answer all the questions accurately, and it would be impossible to differentiate one responder from the other. The computer would seem as same as the person. One could postulate from this that, yes, the computer could acquire a form of consciousness.

The dualist would argue that this is nothing more than "parrot talk" and that the computer had no real understanding of what it was doing. But the singularist would point out that, should the machine be sophisticated enough, might it not develop a sense of self and, perhaps, other more emotional features that we would call human?

THE iPod

Most cerebral scientists would agree that there are about 100 billion to 150 billion neurons in the human brain (however, more recent studies suggest only 85 billion). This comes to about 70,000 cells per gram of brain tissue. Not too long ago, I purchased an iPod with about 150 gigabytes of storage—that is, 150 billion storage cells. Although small and much lighter than the human brain (we seem to have exceeded brain density, in our quest for smaller and more powerful devices), the storage capability would appear to be the same as my brain. Some would argue that the brain has much higher interconnectivity. That may be true, but interconnectivity only goes so far. There is a mathematical limit, and computers are getting closer every eighteen months (according to Moore's Law). Nonetheless, I am convinced (as you are) that my iPod has neither consciousness nor any individual thought.

It would appear that as computer science becomes more sophisticated, there is a growing appreciation for human consciousness, a growing understanding that human thought is obviously more than just neurons.

MICROTUBULES

Stuart Hameroff is a well-known cerebral scientist who has thought about this dichotomy for many years. Hameroff is a medical doctor, anesthesiologist, and professor at the University of Arizona's Department of Anesthesiology and Psychology. He is also the associate director for the Center of Consciousness Studies at that same university. Hameroff recognized that neurons alone cannot explain the complex processes and overwhelmingly complicated mechanisms involved in human consciousness. There just aren't enough neurons.

Hameroff wrote about this in his first book, *Ultimate Computing* (1987). He postulated that intra-neuronal (intracellular) microtubules might be the network that forms the complex function of human consciousness. Microtubules are a complex network of molecular tubes that reside in the brain cell. We really don't know why they are there or what their function is, but there sure are a lot of them, for every neuron has thousands of microtubules. Hameroff used to study this tube network in cell division and speculated that some highly complex computational mechanism must be involved in their development and function. Thus microtubules may be the foundation for human consciousness.

QUANTUM MECHANICS

Two year later, the famous British physicist Roger Penrose published his book *The Emperor's New Mind*. Penrose, a close colleague of Stephen Hawking (perhaps the brightest mind in the world), believed that the mechanism for human consciousness followed the principles of quantum mechanics, or "quantum wave reduction," and thus human consciousness, being nonalgorithmic, cannot be reduced to the simple Turing test. He later called it *objective reduction* (OR) and said that this process allows human consciousness to be reduced to this worldly existence of space-time geometry. This begs the question: What *is* human consciousness, if it has to be reduced to space-time geometry?

Both of these theories were widely attacked by other neuroscientists and physicists, but the question of human consciousness remains. Most physicists believe that microtubules could exist in a quantum state for only a very brief period of time (estimated at 0.00000000000001 seconds), making them useless as a mechanism for human consciousness.

Nonetheless, the puzzle remains. If neurons or microtubules are not the source for human consciousness, then what is? Is it possible that the origin is a quantum mechanical, nonalgorithmic, weightless, massless form of intelligent energy that is somehow integrated with this physical brain? If so, for what purpose and to what end? To find the evidence, we need to address the neuroscience orthodoxy of the near-death experience.

Here are the four main premises of the current academic/orthodox view of the near-death experience.

1. "Had the individual been truly dead, he or she would not have returned."

   Response: The permanent destruction of the brain does not enter into the equation. The relevant fact is that there is a form of consciousness during total global cerebral shutdown, suggesting consciousness may be more than

neuronal. The term *near-death experience* is
a misnomer. In order to have one, the brain
needs to be completely, not nearly, functionless
(dead). Thus these experiences should be called
death experiences (Melvin Morris, MD, says so
in *Parting Visions*, 1994).

2. "Since memory is processed through neurons, a
temporarily dead person cannot remember a near-
death experience. The neurons were unable to
function and thus unable to hold memory at the
time."

Response: If there are two forms of consciousness, as proposed in dualistic theory, one would
carry over while the other shuts down, allowing for a transfer of some memories acquired.

3. "A personal conviction of the reality of the experience does not invalidate a neuropathophysiological
explanation."

Response: Correct, but it also does not negate
the experience. The almost-universal reports of
realistic conviction should not be ignored.

4. "Much of the current literature on the subject of
duality, soul, and near death is limited to touchy-
feely or hearsay evidence by persons of question-
able intelligence, credentials, or motives. Books

on these issues have been created and circulated mainly for the simpleminded and often purely for profit."

Response: Perhaps true. However, one does not throw out the baby with the bathwater.

Thus the orthodox neuroscientific community has, at least in the past, been confined to explaining these near-death "hallucinations" as by-products of a neuropathophysiologic mechanism of the cerebral death process, the physical response to a sick and dying brain. These are some of the theories.

SENSORY DEPRIVATION THEORY

This theory was introduced by Ron Siegal in 1980 and postulates that when the brain is without any meaningful input, stored perceptions are released and become organized into hallucinations. Because any experiment demonstrating this phenomenon would require a normal subject and a normally functioning brain (not near death), this theory has never been taken too seriously among neurologists. One could conclude that the line of thought is illogical.

DEPERSONALIZATION/DEREALIZATION

This theory stipulates that the NDE is an adaptive mechanism produced by the brain to reduce panic and mercifully allow the panic-stricken person to calm down and accept fate. But because nature is functional and purposeful, this theory also has never had a significant following. There is no profit in nature for this mechanism.

EGO REGRESSION

This theory is based on the belief that because of the massive shutdown of the verbal form of consciousness, which the individual has become accustomed to for the majority of his or her life, the ego is repressed to a pre-verbal stage, which is interpreted as mystical. This one went out with Freud and, being unscientific and highly speculative, warrants no further comment.

BIRTH MEMORIES

This theory is rather unique. It is based on the belief that because of the global cerebral shutdown of all memory during near death, only the memory of birth is retained, and it projects into the limited consciousness and is interpreted mystically. When Grof and Halifax described this theory in 1977, they felt that the near-death experience should be called the "near-birth" experience.

I have often considered this theory as proof that one can publish and get high at the same time. I should have such a job...

For perhaps painfully obvious reasons, none of these theories was taken seriously by the neuroscientific community. However, one theory was unofficially accepted as neurological orthodoxy and has since constituted the party line of neurological academics.

TEMPORAL LOBE
EPILEPSY THEORY

Temporal lobe epilepsy theory is probably the oldest and most widely accepted theory of the near-death phenomenon in neurological academics. This is in part due to the fact that the temporal lobe contains Ammon's horn, which is the most epileptogenic area of the brain.

The theory is straightforward. Because of global cerebral shutdown, the brain's seizure threshold is reduced—i.e., near-death patients are more likely to have seizures. The temporal lobe, being the most epileptogenic lobe of the brain, is the most likely site of that seizure. So as the brain swells due to lack of oxygen, the brain is forced downward and pressure is put on the mesio (inner) portion of the temporal lobe, which includes Ammon's horn. And because the temporal lobe is the psychic area of the brain, a focal seizure in this area produces a psychic response. Thus, the chain of events, according to the party line, is cardiopulmonary arrest, anoxic encephalopathy, and seizure resulting in the psychic phenomenon known as NDE. (A persistence of blood perfusion to the occipital cortex, resulting in a persistence of central vision sensation, was a popular explanation for the "light at the end of the tunnel" phenomenon.)

WILDER PENFIELD

This theory was first described by Wilder Penfield, who has often been called the father of American neurosurgery. He was educated at Princeton and Oxford and later at Johns Hopkins University. Dr. Penfield embarked on an ambitious journey by mapping out areas of specific cortical function in the human brain. He used a novel technique of administering microelectric stimulation to specific areas of the brain—frontal and temporal primarily—during craniotomy (brain surgery).

The reason this is possible is because of the fact that the brain has no sensory nerve fibers—that is, no sense of pain. So brain surgery (often to remove a tumor or portion of the brain) was performed while the patient was awake, producing a drug-resistant seizure. The patients were therefore able to describe their experiences while Penfield and crew stimulated different areas of their brains. Without question, the most productive area of stimulation was the temporal lobe.

After more than two decades of this type of research, Penfield published his landmark book, *Speech and Brain Mechanisms*, in 1959. He was able to reduce the psychical responses to two different types: *experiential hallucinations* and *interpretive illusions*.

Experiential hallucinations occurred when a patient seemed to relive a previous event in his or her lifetime. The patient would not only see and hear the recalled event but reexperience it "as though the stream of

consciousness were flowing again as it did in the past."
Interestingly, despite the reality of the hallucination, patients always knew they were in surgery, describing the
events to the doctor and never interacting with anyone
in the hallucinations. The patients always understood
that these images, although vivid, were a reproduction
from their personal past.

I think the most interesting discovery of this research was the fact that the relived memories were of
no particular importance to the patient. This is where
that notion that "you only use 10 percent of your brain"
came from. I suspect that, according to this study, only 1
percent would be more appropriate. This study suggests
that perhaps everything in our personal past is recorded
in our brain, and the problem is not storage but random
access. However, with only 150 billion nerve cells (and
more recent studies suggest 85 billion), how is it stored?
We still have much to learn.

Interpretive illusions are subconscious conclusions
about the present. These are illusions of time and space. A
patient might think that she had the same operation before
(déjà vu) or that everything is strange or unfamiliar (jamais
vu). I recall a patient of mine who suffered a stroke in the
posterior temporal parietal lobe of the left hemisphere of
the brain and thought that everything was new and strange
to him, even in his own house. Sometimes these patients
would describe a sensation of coming out of their bodies or
coming out of this world. It must be pointed out that this
was always a sensation and never an actual coming out and

seeing the world from a different perspective, which is how the near-death phenomenon is often described. All of these Penfield interpretive illusions were just that—illusions—and the patients knew it. This is a far cry from the very real experience of the NDE.

Because of these experiments, most neurologists, including myself, believed that the NDE was an illusion produced by an encephalopathic (sick) brain and nothing more. The uneducated and more simpleminded might be duped into thinking that it was a window into another form of reality, but not the educated, not the informed, oh no, not the academically enlightened. *Not me.* Or so I thought.

Penfield thought this way, and at his farm in Canada there was a large rock on which he painted on one side the Greek symbol for spirit and on the other side the figure of a human head with a question mark inside. He then connected these two images with a solid line linked with the Aesculapian torch (the symbol of medical science). When questioned about these drawings, he would give his standard line: "The brain can explain the mind fully." Penfield's rock symbolized the modern view of the near-death experience. If you wanted to learn about the near-death phenomenon, then you would do better to study medical science and not the patient who had the experience. There is no room for a soul or spirit in this theory. You are your brain, and that is all.

I, on the other hand, had come to believe that the near-death *experiencer* should be listened to and investigated. The true path to knowledge requires open-mindedness and sometimes starting from nothing, without preconceived notions. Where would we be without that approach? No Galileo, no Isaac Newton, no Albert Einstein. Most, if not all, of our great scientific achievements have come from thinking outside the box.

> **"One thing only I know,
> and that is I know nothing."**
> –Socrates, in response to the Oracle of Delphi,
> who called him the wisest of the Greeks (circa 390 BC)

As Penfield grew older and, I believe, a little wiser, he began to question his philosophy of neurological singularity. After several decades of research and thought on the subject, he reversed his position. He symbolized this change in thought by going to that same rock, painting a dotted line between the two symbols, and adding a question mark above this line. He later wrote that "the mind is not something to be reduced to brain mechanisms" and "the brain has *not* explained the mind fully." He also said, "The dualist hypothesis seems to be the more reasonable of the explanations."

Thus Dr. Wilder Penfield, the father of American neurosurgery, the originator of the modern hypothesis of the near-death experience, stopped believing his own theory. So why are we, as neurologists, espousing

the virtues of this explanation when even its originator recanted? It was as if Billy Graham cast off the cloak
of Christianity and became a Muslim and yet was still
quoted as gospel.

Interestingly, this nasty little secret of Penfield's
reversal was not told in medical school. We were fed the
gospel of this mechanism, but the author's "backslide
from the faith" was covered up.

Penfield had good reason to backslide, as the reasons why temporal lobe epilepsy does not hold water
are many.

FIVE REASONS WHY THE PENFIELD MODEL DOES NOT WORK

1.   There is no data suggesting a correlation between the incidence of clinical seizure and a near-
death experience in a near-death event. In fact,
anecdotal experience/evidence suggests there is
no correlation. In all my extensive experience as
a practicing neurologist, I have never seen it, nor
have my colleagues.

2.   A focal temporal seizure producing a visual hallucination is typically a recall phenomenon (experiential)—
a past memory—and not the completely new and
unfamiliar vista and interaction typically seen in the
near-death experience. I have had many patients with
temporal lobe visual hallucination seizures, and none
of them were in any way similar to an NDE.

3.   The interpretive responses described by Penfield were a *sensation* of being drawn out of the body but not actually doing it. The near-death experiencer will describe actually coming out of his or her body and often remark on the difference of seeing oneself for the first time in real time. The patient often says, "Do I really look like that?" This only makes sense when you realize that we never actually see ourselves as others do. We often only see ourselves in the mirror (reversed) or on printed copy (2-D). Thus, the NDE is a real 3-D experience, not Mickey Mouse, Donald Duck, or spiders on the wall.

4.   Ammon's horn is right next to the anterior commissure, which connects the two hemispheres of the brain. Any seizure here would quickly go through the anterior commissure to the other side of the brain, causing a general seizure and, thus, no hallucination. This is, in fact, what usually happens: seizures following a cardiopulmonary event are typically generalized (whole brain) and not focal, as would be required if a visual hallucination were to occur.

5.   Patients who display a seizure as a result of cardiopulmonary arrest typically do not regain consciousness and usually expire. I will explain the reason later. Although Patient 1 (whose brother had the bejeweled sailboat) had a seizure as well as cardiopulmonary arrest, his seizure came first (before the cardiac arrest),

and for this reason he lived to tell his story. The brain had a chance to recoup.

Although many neurologists still believe in the Penfield model, it is no longer at the forefront, and other models are presenting themselves. Almost all of these models are based on the concept of a chemical (neurotransmitter) or ion that stimulates a neuron during near death and causes the hallucination. These molecules or ions are able to fit into a nerve-receptor site (a specific area on the nerve) and thus stimulate the nerve, resulting in a psychic response. There are many different receptor sites in the brain that are sensitive to a specific chemical or ion. These are often referred to as *docking sites*, where the molecule or ion can rest on the nerve and stimulate that nerve to perform its function.

Various theories using these phenomena have been applied to NDE, and I will debunk them all.

ENDORPHIN THEORY

One of the more popular theories is the endorphin model. Glutamate is a neurotransmitter used by a great portion of the human brain. It is a sort of gas pedal or volume control: if pressed too hard or turned up too high, the result would be massive engine failure or a blown amplifier. So if the exposure to glutamate is too excessive, the neurotransmitters become overstimulated and die. Rothman in 1984 revealed that there is a flood of glutamate in the brain as a result of cerebral anoxia or ischemia.

In 1981 and later in 1989, Carr postulated that during anoxia (cerebral vascular shutdown), a flood of glutamate is released into the brain. Endorphins, which are chemical substances theoretically produced by the brain during ischemia, may have a neuroprotective effect by attaching to an opioid receptor site (like morphine) on the nerve and acting like a decelerator or volume control, increasing the chance for survival. Carr thought that this model might explain the "high" often experienced by near-death experiencers.

A similar model is the NMDA (N-methyl-D-aspartate) receptor site theory. This neuron receptor site has multiple docking sites (i.e., it can be stimulated by different chemicals) and is responsible for the hallucinations produced by drugs like PCP (angel dust). There are two neurotransmitters of interest in this model. The first is endorphins, which is a natural substance produced by the brain and which has a euphoric effect. A "jogger's high" is a good example of how this works. The jogger runs, the brain produces the "drug," and the person gets high, reinforcing the behavior of jogging.

Another chemical is endopsychosin. Endopsychosins are a purely theoretical chemical (never identified) that are supposedly produced during brain ischemia or cerebral anoxia.

The reasons why these theories fail to explain the near-death phenomenon are many. I will list a few.

1. There is no experimental evidence that there is a flood of endorphins during global cerebral ischemia, and thus there is no scientific/clinical evidence for this model.

2. Although endorphin stimulation may explain the euphoric sensation of the near-death experience, endorphins are not hallucinogenic and therefore do not explain the visual and auditory effects of the near-death experience. Some researchers (such as Musachio, in 1990) have hypothesized that the endorphin level gets so high as to completely saturate the NMDA receptor site, which has a limited hallucinogenic effect. However, this has not been verified experimentally. Furthermore, any hallucinating effect experimentally noted from opioid receptor stimulation was very limited and most often related to synthetic drugs, not natural endorphins.

3. There is no evidence that endorphins have any kind of protective effect on cerebral neurons. Therefore, there is no benefit for nature to devise such a scheme.

4. In 1980, Oyama discovered that, after injecting beta-endorphin directly into the cerebrospinal fluid of humans, there was no hallucinogenic effect and the analgesic effect lasted all day. This is not the typical brief episode characteristic of an NDE.

So this model is not acceptable as a physical explanation for the near-death experience. Thus, another is required, and so we come to the ketamine theory.

KETAMINE THEORY

Ketamine hydrochloride, also known as Ketolar, was developed in the 1960s and has been used extensively as a *dissociative anesthetic*. Although it was popular as a general anesthetic for children years ago, it is not used for neurosurgery, because it increases intracranial pressure. Ketamine is called a dissociative anesthetic because of its novel effect of having patients feel that they are dissociated from their environment. Often they describe feeling as if they were coming out of their bodies. No visual effects accompany this; there is only the impression or illusion. It causes this effect by stimulating the PCP portion of the NMDA neuron receptor site. (It was also a popular street drug, known as "special K.")

This theory was given further support by the fact that Rothman in 1987 reported a neuroprotective effect of ketamine on ischemic neurons in vitro (in the lab and not in real life). If it were possible that the brain produced ketamine during near death (global cerebral ischemia), then we might have a mechanism for the near-death phenomenon and a reason for nature to use it. However, there are several reasons for this theory to be blown out of the water.

1. Ketamine is a synthetic substance; it is not produced by the brain. It is only created in the lab. Endogenous (naturally coming

from the brain) hallucinogens are called *endopsychosins*, and no one has proven their existence unequivocally.

2. The idea of a flood of ketamine-like substances saturating the brain after cerebral ischemia is not consistent with clinical neurological common sense. If this were true, then patients having strokes would often describe hallucinations and illusions, and this is in no way typical of a stroke. Neither I nor any of my colleagues have ever seen anything remotely similar to a near-death experience from a stroke patient. It just does not happen. If this theory had any basis in reality, stroke patients would be hallucinating like crazy. But they are not.

3. The apparent neuroprotective effect of ketamine is found only in the lab (in vitro) and not live (in vivo). Any clinical researcher will tell you that there is a big difference between vitro and vivo.

The fact is that none of these models work. They don't work because under close scrutiny they don't make sense. The common premise of these theories is that either the cerebral neuron at near death is hyperexcitable and prone to focal seizures (which it is not) or that the receptor sites are stimulated (perhaps for neuroprotection), causing the neuron to create the hallucination of the near-death experience (which cannot be done in the setting of anoxia). When we examine the neuron in near death, we will find that it is not *hyper*excitable but rather *hypo*excitable (sedated), and receptor sites are not functioning at all, because the nerve is only interested in survival and thus shuts down. In other words, there is compelling evidence that during near death, the neurons are not transmitting anything. So if a patient reports a form of consciousness when the neurons are not firing, this is both very relevant and extremely interesting to anyone studying the nature of human consciousness.

Cerebral Ischemia
The Science of Nerve-Cell Death and Dysfunction

The biochemical mechanism of cerebral ischemia (neuron death) has been well known since extensive research on this subject was done in the 1960s. The fuels that keep the brain functioning are glucose and oxygen. Without both of these, the neuron goes through a rapid change in metabolism, shuts down, and soon dies. The brain does not store any glucose for immediate use in the brain. Thus, when the brain is deprived of glucose or oxygen, it goes through a rapid chain of metabolic events that are well understood.

Glucose is necessary for the production of ATP (adenosine triphosphate), nature's universal true biofuel, which the brain truly needs. It's like each neuron has a tiny refinery for converting common crude oil (glucose) into gasoline (ATP). It is ATP that keeps the neuron cell alive, and without it the mechanism that sustains the cell fails quickly.

ATP is made when glucose and oxygen combine with the ever-abundant ADP (adenosine diphosphate). Because oxygen is necessary, this chemical reaction (metabolism) is called *aerobic*. The reaction goes as follows.

$$\text{pry = glucose, P = phosphate}$$
$$2\text{pry} + 38\text{P} + 38\text{ADP} + 6\text{O2} = 6\text{O2} + 6\text{H2O} + 38\text{ATP}$$

Notice that the equation is very efficient. Only two molecules of glucose (pry) can result in thirty-eight molecules of ADP. This is a ratio of 1:19.

If the brain is deprived of oxygen, it can still function on the efficient aerobic metabolism, but for only about fourteen seconds. After this, the brain will have no alternative but to default to *anaerobic* (no oxygen) metabolism.

$$\text{Pry = glucose}$$
$$\text{Pry} + 2\text{ADP} = 2\text{ATP}$$

Notice that this is much less efficient. One glucose molecule creates only two ADP molecules, a 1:2 ratio. Thus, anaerobic metabolism is only one-eighteenth as efficient as aerobic metabolism, and so the cell, in order to stay alive, will have to go through a hell of a lot of glucose—which, of course, will soon run out, because the brain stores only enough for its immediate use.

Interestingly, in 1963, Geiger showed that if oxygen is maintained when glucose stores have been long depleted, neurons can maintain a state of suspended animation for up to one hour. This explains how diabetics with extremely low blood-glucose levels can survive coma.

The problem with the near-death experience is that it is classically a *cardiopulmonary* event. In other words, both oxygen *and* glucose have been cut off, and the brain much more rapidly goes into a state of suspended animation where neurons are no longer functioning. Thus the near-death event is an event of massive and almost instantaneous shutdown of the neuron. This is not an encephalopathic (sick) brain, where it may be a little low on oxygen or glucose—a sputtering airplane or a flickering light. No, it is a sudden and catastrophic shutdown of cerebral function. The airplane is dropping from the sky. The light bulb is out.

There is no cerebral hallucination if the brain is shut down. There is no hallucination if neurons are hyposensitive, and that is just what they are: *hypo*sensitive (i.e., not sensitive). But we have been told that they are the opposite. Now we know that was just bullshit. (It seems that the older I become, the more I like that word, *bullshit* (and I apologize for my somewhat offensive use of the word.) It is true that we have to unlearn all the ignorance of our past if we are to gain any truth in our earthly experience.

Let me explain why the neurons are hyposensitive in the near-death event, by looking at the five stages of the neuron cell at cardiopulmonary arrest.

Stage One. The first stage begins with cardiopulmonary arrest. The lungs and heart are not working. There is no further delivery of glucose and oxygen to the brain. The brain continues to function normally on its own meager stores of glucose and oxygen.

Stage Two. After about fourteen seconds, there is a massive transformation of intraneuronal metabolism, from the normal aerobic to the much less efficient anaerobic metabolism, as oxygen is quickly exhausted and ADP is depleted. The cell starts looking for other sources of phosphate to convert to ATP. It does this by breaking down phosphate bonds often used for signal transmission. (The brain is not interested in signal transmission at this point, only survival.) There are pumps on the neuron membrane walls that must constantly pump out sodium and pump in potassium. Our neurons are like a leaky boat that requires constant use of the bilge pumps. If the pumps fail, then the boat sinks.

Stage Three. This stage begins after about one minute of cardiopulmonary arrest. The stores of phosphate bonds are becoming depleted, and the energy-dependent membrane pumps are beginning to fail. As the pumps fail, sodium begins to leak into the neuron cell, and potassium now is exiting. The excess sodium and the depletion of phosphate bonds within the cell and membrane depolarize the neuron. There is no longer any intracellular machinery or membrane machinery for electrical transmission. The neuron can no longer transmit any information. It is, for all practical purposes, as dead as Julius Caesar. (This electrographic effect was first described by Heyman as long ago as 1964, when he experimented on dogs and monkeys.)

Stage Four. This is the stage that is of much interest to near-death researchers. The brain is now at a level of neurological suspended animation. The brain's only concern is survival. The neurons cannot and do not transmit. Rather than hypersensitive, the neuron is clearly not sensitive at all. It is no more sensitive than a telephone pole. If the cell does transmit any signal, it will deplete any meager stores of ATP remaining, resulting in total membrane pump failure (stage five), and the cell quickly expires, never to return. Thus, the neuron refuses to work; its interest is only survival. It goes on strike until wages are better.

Stage Five. With the complete depletion of ATP, the pumps totally fail. Sodium rushes into the cell, and potassium rushes out. The cell then explodes, the sudden increase in sodium osmotic pressure causing the membranes to burst. There is no return from this phase. The other four stages exist only to prevent this stage.

Thus, the neurons are *not* just hyposensitive. They are *not even working*. No firing neurons and no functioning neuron membrane receptor sites. There is no "neurological inhibition," as promulgated by Susan Blackmore (PhD psychologist in Great Britain).

What is unequivocal and inescapable is the cold, hard fact that a form of consciousness can exist while there is no neuronal function whatsoever. The brain is flat-lined. The brain is just as nonfunctioning as if vaporized by a direct head shot from an 88 mm cannon. Yet there is compelling evidence for consciousness.

The scientific perspective today is clearly moving away from the neuronal origin of human consciousness, as discussed previously. Scientific leaders as far back as Penfield, and now including Roger Penrose and Stuart Hameroff, are embracing an "out of the box" extra-neuronal origin for human consciousness on scientific grounds. I believe that one day this subject will be basic training in our medical schools, and doctors will be better trained because of it.

Understanding that the origin of thought is not the brain is critical, for it opens the door to so many other possibilities. There is the physical and the extraphysical. The extraphysical is just as real as the physical.

I recall asking Dannion Brinkley, who was dead for over twenty-three minutes and who wrote two books and two movies about his story, "What is it really like being dead?" His response: "You'll never be more dead than you are now."

You are a powerful spiritual being having a temporary four-dimensional earthly experience. Embrace it. Learn from it. And don't be afraid.

One thing is certain: if there is an extraneuronal origin to human consciousness, if there is a spiritual side to our existence (and I believe there is), and if ancient aliens existed and still exist today (and I believe they do), then they would have a much better understanding of the two worlds (physical and extraphysical) and how to communicate between them. I believe spiritual growth goes hand in hand with the intellectual. If we, with our meager technology, are beginning to grasp this, then logically they would have had a long-standing intimate relationship with this dimension of existence. And for them, knowledge would have replaced faith, making religion unnecessary. Understanding this will help us understand their motive.

CHAPTER **FOUR**

REINCARNATION
The Evolution of Self

It is no more surprising to be born twice than it is once.

–Voltaire, eighteenth-century French writer and philosopher

Our bodies must have existed without bodies before they were in the form of man, and must have had intelligence.

–Plato, from the book *Phaedo*, quoting Socrates (circa 399 BC)

I was always bold in the pursuit of knowledge, never fearing to follow the truth and reason to whatever result they led.

–Thomas Jefferson (circa 1812)

Once it is understood that the origin of thought is outside the brain, it is not a great leap of faith to come to grips with the idea that a single consciousness may inhabit more than one physical entity. All the great Greek philosophers understood that if duality is indeed true, then the possibilities for existence become almost endless.

In the third chapter, we looked to the science of the near-death experience for evidence of duality. In this chapter, we will look to reincarnation. For not only is reincarnation likely if a soul exists, but a soul (duality) is likely if reincarnation exists. This perspective has been used by ancient Buddhist priests (in search of the next Dalai Lama), by Hindus, and also by the Greeks.

When attempting to use data to support reincarnation, the data must be *veridical*. This means that there is one and only one explanation: the data could have not been obtained any way other than by living a previous life. You recall the story from chapter three about the Dutch gentleman who, while in a coma from cardiopulmonary arrest, was able to see the nurse put his dentures in the crash cart drawer. This is veridical data.

HYPNOSIS

There are two primary methods for investigating reincarnation. The first is through *hypnosis*. Hypnosis is an actual neurophysiological phenomenon in which the patient is asleep but the EEG demonstrates an alpha (eight cycles per second) or "awake" rhythm. (The EEG for sleepwalkers, on the other hand, is a deep *delta* rhythm of sleep—four cycles per second.) This schism is rarely achieved, but when it is, you can get some interesting results.

The problem is you need to be a good hypnotist (rare) and have a good subject that is easily hypnotized (very rare). Most hypnotists are pseudohypnotists, and the hypnotic state achieved is a shallow state of pseudohypnosis (not the real deal). To be truly hypnotized, one needs to be asleep and have an alpha pattern on the EEG. Clearly, there must be an EEG present for a true hypnotic session. The second is through the interviewing of children, as, in this early stage of life, they are closest to the spiritual realm, from whence they came. Let us first examine the hypnotic method.

THE CASE OF VIRGINIA TIGHE

Virginia Tighe was one of those people who was very susceptible to obtaining a hypnotic state. Reportedly, she had no preconceived notions about reincarnation, and her personal history was rather typical of a woman of her day. She was born in Madison, Wisconsin, in 1923 and moved to Chicago with her family at the age of three. She attended Northwestern University for only a year, and then she married. Unfortunately, her new husband died in 1944 during World War II. She then married a businessman, Hugh Brian Tighe, moved to Colorado, and had three children with him.

This case was reported by Morey Bernstein, who was a hypnotist. While under hypnosis, Mrs. Tighe revealed a past life: Bridey Murphy, a nineteenth-century Irish woman. After six hypnotic past-life regressions, Bernstein wrote his famous book *The Search for Bridey Murphy* (1956).

While under a deep hypnotic trance, Virginia became Bridey Murphy, stating that she was born in Cork (a small coastal town in southern Ireland) in 1798. She said she had a brother, Duncan Blaine Murphy, who was named after her father, Duncan Murphy. She reported that her mother's name was Kathleen. Although her family was Protestant, she married a Catholic, John Brian Joseph McCarthy, and moved to Belfast, where John attended law school and later taught at Queen's University. She died childless in 1864, at the age of sixty-six.

During the 1950s, many newspaper and magazine articles either proved or debunked her story, but many people (including me) who have personally listened to the recorded tapes of her hypnotic sessions are convinced of her sincerity. Interestingly, it was these debunkers who eventually helped verify the fact that Virginia Tighe may have indeed been the reincarnation of Bridey Murphy. Furthermore, it was unlikely that Virginia lied for financial reasons, as she did not profit whatsoever from her story.

1. Bridey gave the exact names of the only two grocers in Belfast at the time (Fars and John Carrigan), as well as

named the rope company and the tobacco house. These names were later verified by the old town records.

2. Bridey later claimed to have read a book entitled The Sorrows of Deirdre. Critics revealed that this book was not published until 1905. However, after further investigation, it was discovered that a popular paperback book entitled The Song of Deirdre and the Death of the Sons of Usnach was published in 1808, when Bridey was eleven years old. Historically the words song and sorrows were often used interchangeably. For example, one might say, "Let me tell you my song."

3. Bridey mentioned that the coinage of the day was the *tuppence*. This was later found to be incredible information, as only rare-coin experts would acknowledge that this coin was used in Ireland from 1797 to 1850.

4. It was also interesting to observe that Bridey seemed to embellish her social position and was most probably of a lower class than she presented. Her husband was probably a courier or clerk to a barrister. This was revealed by her low-class accent, which was almost unintelligible. Theatrical experts recognized and verified her speech. Her dishonesty in this respect, rather than proof of fraud, is actually very human and was likely for someone of her position.

THE ARNALL BLOXAM TAPES

While Bernstein's Virginia Tighe tapes reveal a compelling case for a single past life, the Arnall Bloxam tapes present compelling evidence for multiple past lives. Jane Evans was another person who was easily hypnotized. She presented six different past lives. Her story was told in Jeffery Iverson's book *More Lives Than One* (1976).

Arnall Bloxam was born in 1900 and grew up in Pershore, Worcestershire, England. He served aboard a minesweeper in World War I and later in World War II as a navy lieutenant. Although his real dream was to become a medical doctor, he developed typhoid fever and, from fear of infecting others, gave up on his medical career. He settled in Cardiff in South Wales and developed a reputation as a good hypnotherapist. Each week, Bloxam would invite guests to his house and hypnotize them, and he'd often tape-record the visits. He developed quite an extensive file of recordings, but the most impressive were those of Jane Evans.

Jane was born in 1939. After repeat regressions under hypnosis, she consistently revealed six past lives spanning almost two thousand years, starting in ancient Roman England (circa AD 286) and ending as a nun in the midwestern United States (circa 1870).

FIRST LIFE: JANE THE ROMAN WIFE

The oldest of Jane's past lives was in ancient Roman Britain around AD 286. Her name at the time was

Livonia, and she was the wife of a teacher of Latin, Greek, and poetry, a certain Titus. He was employed by a wealthy Roman named Constantius, who was married to Helena and lived in a beautiful villa named Eboracum in Roman York. Titus was the tutor to Constantius's son, Constantine.

Livonia described an interesting event in her life. It involved a man named Allectus, who visited her employer's house with a summons demanding that Constantius return to Rome. She then explained that the Roman Empire had been split in two, with Diocletian controlling the east and Maximianus controlling the west, which, of course, included Britain. Livonia also revealed that she did not like this Allectus, for two reasons: he had cold eyes, and he visited a man named Carausius at the port of Gassoriacum (currently named Boulogne, in France) just before coming to Britain. This apparently raised the index of suspicion of a traitorous conspiracy, for this Carausius was a powerful and ambitious man who controlled the Roman fleet in Gassoriacum.

Being summoned to Rome, Constantius left his wife in Britain. He later divorced his wife in the temple Jupiter on the Palatine in Rome, which was a shrewd move, paving the way for his more political marriage to Princess Theodora, daughter of Maximianus. He subsequently had two daughters by her.

In further hypnotic sessions, Livonia's fears were confirmed. There was a power grab orchestrated by

Allectus and Carausius, forcing her to flee to Eboracum. She later revealed that Allectus murdered Carausius and declared himself ruler in Britain. Later she reported that Constantius returned to Britain with a large army, executed Allectus, and put the rebellion to rest.

Livonia described Helena's sadness, learning of her husband's divorce and remarriage. Helena then went to Verulamium (currently St. Albans, Hertfordshire, England), met a wood carver named Albanus, and became Christian. Sadly, they were both killed in the Christian purge initiated by Galerius and sanctioned by Diocletian (circa AD 305).

This case is relevant for several reasons.

1. The names, places, and events are all true to historical record and are known only to experts in ancient Roman history, which Jane was not.

2. There are common coins of Diocletian and rare coins with the faces of Carausius and Allectus, produced only during their rebellion. Even Constantius had a coin.

3. At no time during the recording was Livonia's political knowledge inappropriately advanced for her time. For example, at no time was she aware that Titus's pupil, Constantine, was to become the first Christian emperor of Rome. Her knowledge was always appropriate for the time.

SECOND LIFE: JANE THE JEWESS

In this past-life regression of Jane Evans, she reported being a twelfth-century Jewess named Rebecca living in a wealthy neighborhood in the northern portion of York, England. Her husband was a wealthy moneylender named "Joseph of the seed of Ezekiel" and was able to prosper under the protection of King Plantagenet (Henry II). Interestingly, she described the recruitment technique used during the Third Crusade. She also described the dishonest dealings of a certain man named Mabelise, who refused to repay his debt to her husband. Sadly, the good King Henry II died, resulting in an uprising of anti-Semitic prejudice and the York massacre of 1190, in which 150 Jews were killed. Although Rebecca fled to St. Mary's Castlegate Cathedral, she, too, was murdered, while hiding in the church crypt.

This recording is relevant for the following reasons:

1. Critics initially discredited this story because, although St. Mary's Castlegate Cathedral is well known, it was never known to have a crypt. However, in 1975, while converting the church to a museum, the crypt, complete with arches and stone vaults from the Roman period, was discovered, confirming Rebecca's historical account.

2. Rebecca always gave an appropriate amount of knowledge surrounding the political developments

of the period on which she reported. Although she was caught up in the political and religiously big-oted frenzy of the time, she was unable to give explanation or insight into the political developments that resulted in her brutal murder.

3. Historical records later revealed that the dishonest man described by Rebecca was no doubt Richard Malebisse, a known anti-Semite at the time and agitator of the 1190 massacre of the Jews. Being a debter to Joseph, he obviously had a profit motive.

THIRD LIFE: JANE THE SERVANT GIRL

In this hypnotic session, Jane described herself as Allison, a servant girl to a certain Jacques Coeur, who lived in the city of Bourges in the south of Paris, circa 1450. She gave a good description of the mansion and stated that it was one of many. Allison was originally from Egypt and was purchased by Coeur while he was on a business trip in Alexandria. She went on to describe many of his possessions, including a golden apple. As evidenced by her speech, Allison was not an educated girl. But, being close to Coeur, she became privy to some of the political intrigue of her day. She identified the current king of France as "Charles de Valois" and described his physical appearance as well as his mistress, Agnes Sorel, whom she stated was a personal friend of her master. She also identified the king's mother as Duchess Yolande, who often visited Coeur on behalf of the king to borrow money. She also stated that the king chose not to live in Paris but rather the "castle at Chinon" nearby.

Again with uncanny accuracy, she identified the major political players of her day, including Louis the Dauphin, who was commonly believed to have murdered (poisoned) his wife, Margaret of Scotland. She also reported that the king borrowed money from her master for the ransom of the king's brother, Rene d'Anjou (actually the king's brother-in-law).

She also reported that the king's mistress, "the fair Agnes Sorel," died. Poison was suspected, and Louis de Valois started the rumor that her master did it.

Coeur was brought to trial, and because so many people owed him money, including the judges, he was conveniently found guilty, arrested, and his fortune confiscated. Allison, alone and an infidel in fifteenth-century France, facing a ghastly burning at the stake, poisoned herself and died.

Allison gave surprising insight into little-known historical facts of the time.

1. She accurately described the physical appearance of the king and correctly named him as "Charles de Valois," which is the family name of Charles VII. She also stated that he did not live in Paris but in a nearby "castle at Chinon." Although the king had two castles in the general area, called

Mehun-sur-Yevre, a servant girl may not be expected to know this.

2. She spoke of the king's mother and how she came to her master to borrow money for the king, and she named her correctly as the Duchess of Yolande. This is historically accurate, although she was actually the mother-in-law, something a servant girl may not know3 She also mentioned Louis the Dauphin and the belief that he murdered his wife, Margaret of Scotland, via poison. The historical record never proved this murder, but the rumor was indeed popular at the time. Allison never stepped outside her time in history. For example, she never mentioned the fact that Louis the Dauphin went on to become King Louis XI of France, as this was something that happened after her death.

3. Coeur was a real figure in history. We know Allison's description of his house in Bourges is very accurate, for it still stands today. Coeur was wealthy from a license granted by the pope to import gold, jewelry, and fine silk from the "infidel" Arabs. She also described Coeur's many artistic possessions, including paintings by Jean Fouquet and Jan Van Eyck. She also described "a room where he kept his porcelain and jade and...a beautiful golden apple with jewels in it. He said it

was given to him by the sultan of Turkey." Much of Coeur's possessions, being confiscated by the king when Coeur was tried, became part of the historical record. While Jeffery Iverson was researching his book, a friend with access to the historical record was able to find the obscure list of the confiscated possessions of Coeur, and there on the list was the golden apple.

4. Finally, one did not need to read between the lines to discover that Allison had an adolescent crush on her master. I suspect she was a sort of mistress to him later in life. This type of subconscious footnote makes the impression all the more real.

FOURTH LIFE: JANE THE LADY-IN-WAITING

The next three incarnations were less exciting. Jane recalled being Anna de Castile, a lady-in-waiting to the infant Catherina, daughter of Queen Isabella of Spain, circa 1700. With uncanny detail, she described her sailing trip to England and the events surrounding the marriage of Catherine to Prince Arthur, son of Henry VII of England. Apparently, she died on the return voyage back to Spain.

FIFTH LIFE: JANE THE SEWING GIRL

Jane then described her life as a common teenage sewing girl, Anna Tasker, in London, England, circa 1700.

Although overworked and uneducated, Anna was able to give accurate detail regarding the social and political events of her day. She was able to accurately describe the current monarch as Queen Anne and called her the "fat lady." She knew that the Duke of Marlborough was creating an army to fight the French, because she had two brothers in that army. She also gave a good description of the funeral parade for the Duke of Gloucester and knew that he was the only son of Queen Anne. Anna's life was very hard, and she died young in a dark room with other sick people, possibly from the plague.

Sixth Life: Jane the Nun

This last past life was as Sister Grace, an American nun from Des Moines, Iowa, circa 1870. Her mother's name was Irma and her father's was Clarence. Both died when she was young. Later in life, she moved to a convent in Maryland (possibly the Sisters of Mercy) and gave confession to a Father Ignatius. She was obviously uneducated and not up on current events. In old age, she became crippled with arthritis and died around 1920.

Thus, Jane had a varied and extensive multigenerational experience. That all of her lives were female and of the same race (with the possible exception of the Egyptian servant girl) suggests a deeper extraneuronal origin to gender.

At any rate, her story begs the question of divine purpose. Is this the way of karma? Does God "roll the dice" of genetics, and once the experience is varied and extensive enough, after many reincarnations, does the soul move on to a more advanced stage? Or is it all just endless, unorchestrated, and without purpose?

Although most past-life regressions are rather mundane—no great figures of history—there is one Bloxham tape that was of great interest to the leadership of Great Britain in the 1950s (as well as to me, as a sailor).

THE CASE OF THE
PRESSED SEAMAN

This involves the hypnotic regression of Graham Huxtable, who lived in central England. Graham had no contact with the navy or nautical things, had no interest in the sea, and had never been to sea. He served on a tank in World War II. However, under hypnosis, he gave a perfect description of a British sailor in the eighteenth century.

He described his life as a gunner's mate of the ship *Aggie* under the control of a "Cap'n Pearce" on blockade duty off the coast of France, near Calais. During the recording, it becomes painfully obvious that life aboard ship was hard and unpleasant. The food and water were often mixed with worms and weevils. He describes the bread as being hard enough "to take your teeth down."

The descriptions of naval battles were also painfully real. He would bark things like, "Keep your eye down, boy! Keep 'em in the boat!" (Possibly to *powder monkeys*—young boys responsible for bringing the gunpowder to the guns) so as to avoid blindness from the splinters as a result of enemy cannon fire. He described the captain as being "fond of the cat" and laughed when he said that he liked to wear his hat "thwartships." Graham described the bosun's mate as being "free with the starter." He was also able to give a complete description of the loading, setting, and firing of cannons, including the swinging of the match (which was done to prevent it from going out) and the constant adjustment of the chocks for elevation during sighting.

The hypnosis of Graham Huxtable is relevant for the abundant use of salty expressions and unique nomenclature unknown to a land person unaccustomed to nautical things and nautical history. Because of the abundant nautical expressions, the tape was investigated by the former first lord of the admiralty, Earl Mountbatten, who subsequently requested the assistance of a naval historian, Oliver Warner, as well as that of the earl's nephew, Prince Phillip. A few of the discoveries include the following:

1. The word "thwartships" is an archaic illiterate pronunciation of the established nautical term "athwart," which comes from the Greek meaning "to mount across." By laughing when he described his captain wearing his hat "thwartships," he was meaning side to side.

2. "The cat" the captain was "fond of" was the cat-o'-nine-tails used for punishment aboard ship.

3. The bosun's mate (the foreman aboard ship) was said to be "free with his starter," meaning a small rope whip with a knot (a Turk's head) at the end, used for discipline aboard ship.

This is all veridical data, information only acquirable through real experience of a past life and revealed via hypnosis. Anyway, let's discuss the second method of investigating past lives.

INTERVIEWING CHILDREN

Children, being new to this world, are believed to be closest to their previous state of existence, and therefore they are sometimes able to give information about or valuable clues to their past lives. This is the technique used by Tibetan Buddhists to find the next Dalai Lama. The current Dalai Lama, Tenzin Gyatso, is believed to be the reincarnation of the previous Dalai Lama, Thubtan Gyatso, who expired in 1933. Thus, the current Dalai Lama is believed to be the fourteenth reincarnation of Gedun Drub, the first Dalai Lama, who was born in AD 1391.

How is the next Dalai Lama determined? After the death of the contemporary Dalai Lama, the Buddhist monks of the "High Lamas" begin to investigate evidence for a possible reincarnation of the last great religious leader. They visit the holy lake of Lhamo La-tso in central Tibet, meditate, and watch for signs in the lake from the female guardian spirit, Palden Lhamo, who has promised to protect the chain of reincarnation of the Dalai Lama. These signs may be in the form of visions or dreams received through meditation. Once the process is started, children are questioned and given a selection of toys and other items owned by the previous Dalai Lama. The process is very thorough and complicated, but eventually, when a likely child is found, they invite the Living Buddhas from the three great monasteries to help confirm their findings. When the child is confirmed, the central government is informed. The boy and his family are then taken to the Drepung Monastery, where he undergoes training and preparation for his duties as the future Dalai Lama.

Dr. Ian Stevenson, professor of psychiatry and director of the Division of Personality Studies at the University of Virginia, has had a personal interest in reincarnation for decades. You may remember that the university's founder, Thomas Jefferson, mandated that the university maintain an open mind and not be a "slave to dogma." Dr. Stevenson took that philosophy to heart, an act of bravery in the closed-minded and conservative world of his time.

Dr. Stevenson did not trust hypnotism as a research tool for reincarnation. This is because of its commercial exploitation and because the historical information shared while under hypnosis was often not veridical—it could have been obtained through movies or books (cryptomnesia) and not as the result of past-life phenomenon. Stevenson thus elected to use children as his primary tool for investigating reincarnation. Children are often open and honest and do not have sufficient personal history to cloud data with cryptomnesia.

Dr. Stevenson's publications and research are extensive. He accumulated over 2,500 case histories in his files from all over the world. (Many of these cases were not fully examined, due to lack of funds.) His many books include *Twenty Cases Suggestive of Reincarnation* (1960), *Unlearned Language: New Studies in Xenoglossy, Telepathic Impressions: A Report of Thirty-Five New Cases*, and *Children Who Remember Previous Lives*. A more recent book, *Reincarnation and Biology* (1997), includes photographs that correlate birthmarks and birth defects with reincarnation.

Dr. Stevenson passed away in February of 2007. Needless to say, his research has been exhaustive as well as compelling, and any serious study on the subject of reincarnation would be incomplete without a review of his data.

There are many studies documenting reincarnation by way of interviewing children, much of it scientific and performed by the Division of Perceptual Studies (formerly the Division of Personality Studies, Department of Psychiatry and Neurobehavioral Sciences) at the

University of Virginia, in Charlotte. Dr. Ian Stevenson started this department in 1967, dissatisfied with the contemporary dogmas of psychoanalysis (now for the most part debunked) and behaviorism (which simply did not explain enough).

THE CASE OF JAMES LEININGER

The book of James Leininger (*Soul Survivor: The Reincarnation of a World War II Pilot*, 2010) is a *New York Times* bestseller and is highly recommended reading.

Ever since the age of eighteen months, little James had been fascinated by airplanes, especially War World II American fighter aircraft. However, there was a serious if not nerve-racking problem: James would often scream at night, apparently from terrifying nightmares, waking up his parents. "Fire, little man can't get out!" he would scream. He also would draw pictures of a piston fighter airplane falling to the ground in flames and a Japanese fighter plane nearby.

As James Leininger became a little older, he would explain that he was a Corsair pilot on the ship *Natoma* and died in a flaming plane shot down by the Japanese. He also explained that his name at that time was also James, and his best friend was a certain Jack Larson. His father, Bruce Leininger, being a fundamentalist Christian, did not believe in reincarnation and so set out to prove that this was just the wild imagination of a two-year-old.

I don't want to go too much in depth, for I highly recommend reading the book itself. In summary, Bruce discovered that there was indeed an aircraft carrier *USS Natoma Bay* and that there was a James M. Huston shot down by the Japanese on March 3, 1945, during the Battle of Iwo Jima. In addition, there was a certain pilot, also of the *Natoma Bay*, named Jack Larson, who was still alive at the time and was indeed a good friend to James M. Huston.

Critics argued that because James M. Huston was shot down not in a Corsair, as little James had stated, but in an FM-2 Wildcat, this was not a valid recollection of a past life. However, further investigation revealed that James M. Huston was part of an elite squadron (VF-301) of pilots trying to figure out how to take off and land the famous Grumman Corsair from aircraft carriers. After this service of duty was over, he was then transferred to the *Natoma Bay*. Clearly, then, he *was* a Corsair pilot, and any World War II US Navy fighter pilot would be sure to let his comrades know of it. And so little James let us know of it as well. Jack Larson, his sister, and the sister of James M. Huston are all convinced that little James Leininger was the real deal and was indeed the reincarnation of their loved one, James M. Huston.

(An interesting footnote: little James also explained that he *selected* his parents, but that part you will have to read for yourself.)

Conclusion

There is compelling evidence for reincarnation. This should not be too difficult when one realizes that there is indeed a soul and that this intelligent spirit is a separate entity from the physical body it is temporarily using. I often feel like laughing when someone tells me, "I don't believe in a soul or reincarnation, because I am a *rationalist*." I often reply, "So you must consider me an *irrationalist*. I used to have the same frame of mind...when I was ignorant of such things."

Tibetan monks describe the period of the soul's existence between carnal lives as the "in between" period or the "bardou." This is the period when the soul is in preparation for the next destination (be it another physical life or something else). Again, one needs to ask: For what purpose and to what end? And again we must consider that the ancient aliens, who are still among us and who have known the truth about consciousness since humans first appeared, still have much to reveal. Is it possible that reincarnation is a process that is not only familiar with them, but something that they depend upon?

Chapter **Five**

Archeological Enigmas
The Stones Cry Out to Reveal the Past

Know the truth, and the truth shall set you free.
–Jesus (circa AD 30)

The Great Sphinx

Author's photo
The largest and oldest monolithic statue in the world is
perhaps also the most famous archeological structure in

the world. Alone, lying down on the west bank of the Nile at the Giza plateau in Egypt, facing east to the sunrise, the Sphinx is huge. Carved in limestone and measuring over 240 feet long, 20 feet wide, and 66 feet high, it has the body of a lion and the head of a man. Behind the Great Sphinx lie the three Great Pyramids of Giza. It is as if the Sphinx were some sort of guard dog to the pyramid complex.

All Egyptologists will acknowledge that the Sphinx is very old. Mainstream archeologists and academics believe that the Sphinx was built during the reign of the Pharaoh Khafra (2558–2532 BC), who is also purported to have built the second Great Pyramid, which lies behind the Sphinx. How sure are we of this date? The ancient Egyptians tell us nothing of who built it.

Reportedly, the Dream Stele (a large, flat stone with inscriptions) was created and erected by Pharaoh Thutmose IV circa 1390 BC. It mentioned a certain Khaf (not Khafra) in relation to the Sphinx. The extent of that relationship (builder? restorer? worshiper?) and, for that matter, the identity of Khaf are unknown, and unfortunately the stele is now damaged beyond repair. Nonetheless, in the valley temple adjacent to the Sphinx was found a diorite carving of the head of Khafra. It was thus assumed that the head of the sphinx was carved in his image. The two do bear some resemblance, but no official or professional forensic investigation has been performed to confirm this line of thought. It should be noted, though, that in 1993, Frank Domingo, a forensic scientist of the New York police department, reported that the Sphinx head was not a copy of the head of Khafra and did not even depict a person of the same race. So the mystery continues...

THOSE DARN WATERMARKS

Another bit of archeological evidence contrary to the Khafra theory is the Stele Inventory, discovered by Auguste Mariette (founder of the Egyptian Museum in Cairo) in 1857. This stele, built around 550 BC, tells the story of how Khufu (Khafra's father) found the Sphinx buried in the sand. This is perhaps not too surprising, considering the fact that, if left alone as it stands now, the Sphinx would indeed be buried within a few decades, due to windblown sand and shifting dunes. Nonetheless, mainstream academics dismiss this stele story as poetic license or late-period Egyptian revisionism and still cling to the Khafra theory. After all, the Khafra theory is in the books, and academics need their books sold.

R.A. Schwaller de Lubiez, although not a professional archeologist, studied the Sphinx in great detail and reported in the 1950s that because of water erosion noted on the walls of the Sphinx's enclosure, it must be much older than purported by orthodox academia. John Anthony West came to the same conclusion and wondered, "Who is best at dating stones?" He rightly called on a geologist, Robert M. Schoch, associate professor of natural science at the College of General Studies, Boston University. They did an extensive study, including paleometeorological (the effect of climate on stone) examination and stone-core sampling, and they concluded that the rain wear on the body of the Sphinx and its interior stone enclosure was a great deal older than reported by mainstream Egyptologists and most likely dates between 5000 and 7000 BC—a period in history we know little or nothing about. This is because the last period of significant rainfall was before 3500 BC, due to the end of the last glacial epoch. And since Khafra lived around 2500 BC, the Sphinx must have preceded him. In addition, historical evidence suggests that the Sphinx spent most of its time buried in sand and thus was unlikely to suffer much rain erosion from what little rain exposure there was in that period.

I recall reading Robert Schoch's report on the Sphinx to the American Association for the Advancement of Science (of which I am a member) in 1992. I was delighted by what a brilliant idea it was to request a geologist to date weather-carved stone. Although some climatologists may debate the rainfall issues (not many do), Robert's investigation was exhaustive. He pointed out that mud brick structures nearby, much more vulnerable to rain and unequivocally dated to the 2500 BC era, were relatively undamaged, confirming standard climatological conclusions regarding rainfall in Egypt's Giza plateau. In addition, other geologists, such as David Coxill, working independently of Schoch, came to the same conclusion, as reported in the *Journal of Ancient Egypt* in 1998.

I have been to the Sphinx as well as the valley temple nearby, and when viewed in person, it is very clear that the most exposed portion (the head) suffered the least from weathering. This alone should raise questions. In addition, the head is a great deal smaller than the body. It seems almost painfully obvious that the original carved head must have been much larger, perhaps the head of a lion or a jackal (Anubis) which, being exposed, was weathered beyond recognition and subsequently reshaped by later pharaohs. It is interesting to note that there is no evidence there was ever a mane on the sphinx, something a lion would not be without. Also, the jackal or dog Anubis was very special in Egyptian religion, as he was the god who guarded the Earth and the underworld and protected the body of Osirus, the most venerated of Egyptian gods. But the general consensus remains that the body is that of a lion.

Let us digest the few facts we have.

1. We really have no idea by whom or when the Sphinx was built.

2. We know it is old—very old—and dates to well before the fourth Egyptian dynasty, 2500 BC, and most likely dates 7000 BC or before. A period of time we know little or nothing about

3. We know it sits on the western bank of the Nile, facing directly east across the Nile River.

4. We have no idea what the head was when constructed (human, jackal, or lion), but we generally accept that the body is that of a lion.

5. We know that it lies in front of the Great Pyramid complex at Giza.

6. We know that the ancient Egyptians were almost insanely sensitive to the positions of the stars and had a particular interest in the constellation Orion, believing that this was the place of their gods' home and that their god Pharaoh would return there upon his death.

7. We know that the Earth wobbles like a spinning top, but at a slow rate (one cycle every twenty-six thousand years) and that this gives us a different zodiac sign approximately every two thousand years. We are currently ending the age of Pisces and entering the age of Aquarius. Before Pisces was the age of Aries, the ram (2000–0 BC); before this was the age of Taurus, the bull (4000–2000 BC); and so on.

8. The age is determined by the sign (constellation) observed directly to the east during the spring (vernal)

equinox as the sun rises in the early morning hours. This is the time when the sun rises directly in the east.

Let us also consider what the famous Belgian engineer and astronomical scholar Robert Bauval discovered: that the three pyramids at Giza represent, by size and position, the stars of Orion's belt (where the size of the pyramid reflects the intensity of the corresponding star's brightness). Notice that the less intense third star and smaller pyramid is a little offset. We know through Egyptian art that the Nile represents the Milky Way of the night sky. And Robert Bauval found that the position of the three pyramids at Giza relate to the position of the Nile River exactly as the position of Orion's belt related to the Milky Way as seen in the night sky in 10,450 BC.

So let us accept one simple postulate: when originally constructed, the Sphinx was not a sphinx at all but simply a huge carved lion (head included, though it was weathered and replaced over time). If we accept this postulate, then the message of the Giza plateau becomes a little clearer.

What was the zodiac age at this crucial time, in 10,450 BC? Leo, the lion. So this giant statue of a lion, facing the horizon in front of the pyramids, is looking at himself (the sign equates with the sun) rising in the east during the vernal equinox of 10,450 BC.

The facts line up, but whether the conclusion is true is still up to conjecture. Most academics still ignore this perspective and won't even discuss it. A great reference for further reading is Graham Hancock's book *Fingerprints of the Gods* (1995).

Historical scholars and academics would say that this is the archaic period in human history (anything older than 3000 BC), when nothing significant was supposed to have happened. No great developments technical or otherwise. Our ancestors were nothing more than hunter-gatherers. No sophisticated societies. No complex construction. No technology to speak of. But is this true?

GOBEKLI TEPE

In southeast Turkey, at the foothills of the Ararat Mountains, near the town of Sanliurfa, buried under sand are perhaps the most incredible if not the most important archeological discoveries of the modern era. Initially discovered in 1994 by a Kurdish shepherd who noticed a carved stone sticking out of the ground, it is now being excavated by a German archeologist. What he is revealing are scores upon scores of large, perfectly carved T-shaped sandstone pillars with beautifully rendered animals such as geese, cows, lions, monkeys, and wild boar, as well as human figures, finely chiseled in relief and in the round. These stone structures were obviously made with sophisticated tools not known to be in the archaic toolbox.

At the request of the Turkish government, Dr. Klaus Schmidt of the German Archeological Institute has been excavating this site since 1995. Some of these carvings are on display at the nearby Museum of Sanliurfa. The site is huge, and after fifteen years of digging, only 5 percent has been uncovered. Each one of these stone columns stands nineteen feet tall and weighs fifteen tons. They are arranged in circles of roughly ten columns each. Some of these columns were set into bedrock and others in a concrete-like substance. Carbon-14 dating of the organic matter found at the various levels of digging confirm that Gobekli Tepe is about twelve thousand years old.

This was the end of the last ice age, a very dark period of human history that we know nothing about. It's about 7,500 years older than the official age of the Great Pyramid. No tools or agricultural implements have been found. Some experts have suggested that Gobekli Tepe was built and then purposefully buried for preservation around 8000

BC. Preservation from what is, of course, pure speculation. Weather? We know that the end of the last ice age was very turbulent, with flooding and other dramatic climatic changes. Or was it buried for protection from human invaders? Some have speculated that Gobekli Tepe was a monument to the preservation of animal life after the Great Flood.

Whatever the case, this is the oldest culturally sophisticated archeological dig in history, and clearly the history of mankind is going to have to be rewritten. For this site was obviously *not* constructed by primitive cave dwellers using stone "tools" and living on an encounter basis.

Artistic representation

PUMA PUNKU

Puma Punku is part of what appears to be the large temple complex of Tiahuanaco, high (thirteen thousand feet) in the Andes Mountain range, near Lake Titicaca in Bolivia,

South America. There are countless massive, perfectly carved red sandstone structures weighing approximately 120 tons each. Computer analysis of these huge blocks have revealed that each one is identical to the next and can fit into one another other like pieces of a child's Erector Set. Petrochemical analysis indicates that these huge stones were transported up a steep hill from a quarry ten kilometers away. The carved sandstone blocks are so sophisticated in their construction, with exvaginated and invaginated crosses, it would appear that they have been machine-made in a prefabricated manner. One of the stones shows a perfect continuous groove one centimeter wide with drilled holes spaced, with machinelike perfection, equidistant from each other within the groove as if in preparation for the placement of pegs. In addition, the placing of the stones was obviously done with advanced engineering knowledge; the technique today known as *layering and depositing* (layers of sand followed by layers of composite) forms a sturdy base for extremely heavy structures, and this technique was used abundantly at Puma Punku. Investigators often wonder, when does raw human labor stop and higher technology begin? I believe, clearly, it begins at Puma Punku.

But this is not all.

As reported by Arthur Posnansky (1873–1946) as early as 1910, during his exhaustive exploration of the site, he complained that much was being taken away from the area for construction in the nearby city of La Paz. He documented his discoveries with photos and very technical drawings. Some of the stones he captured were fashioned from granite, a stone much harder than sandstone, and some even from diorite. Only diamond is harder than diorite, and we are sure the Aymara (the ancient people of this area) had no diamond power tools. They didn't even have a written language.

But this is not all

These stone structures were apparently clamped together with metal clamps. The Aymara had no metal except for gold, which is unfit for clamping huge stones together, and physical evidence suggests that the metal clamps were of a sophisticated copper-arsenic-nickel-bronze alloy. It looks as though in some spots they were hammered in place, and in others they were heated to above melting point and poured into place.

But this is not all.

Some of the "carved" stone structures had been cut, angled, and grooved in ways that defy explanation. The cuts are so precise that they would be extremely difficult to pull off even with today's technology. They simply must be seen to be appreciated. The only plausible explanation, using today's technological perspective, is that this extremely intricate precision cutting was for the purpose of matrix die casting, to make metal machine parts—machine parts obviously of *no* use to the Aymara Indians.

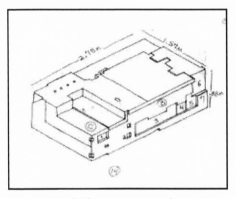

Artistic representation

The dating of this site is all over the chart. Binghamton University anthropology professor W.H. Isbell, using radio-carbon dating, placed mound fill material at approximately AD 440. However, the Bolivian Centro de Investigaciones Arqueologicas in Tiwanaku found matter dating to 1580 BC. Another dating technique, obsidian hydration, gave obsidian objects at the site a date of 2134 BC.

If one takes into consideration the possibility (which there is ample evidence to support) that Puma Punku was a harbor facility on the lake shoreline, we will have to shift our time line back even further, to the end of the last ice age—over 4000 BC—when water levels of the lake were much higher. If one accepts the belief that the Tiahuanaco facility was a huge celestial observation complex, as reported by Dr. Hans Ludendorff and Dr. Arnold Kohlschutter of the German Astrological Commission in 1938, the date may be pushed back to as far as 10,000 BC or more.

If you ask the Aymara Indians who built Puma Punku, they make it very clear that they did not. Their legend on the subject is that Puma Punku was built by giants at the beginning of time. They did not follow the rules of behavior instructed by Viracocha (a Christlike figure of South America, a tall man with white complexion, white hair, and white beard) and were destroyed in the Great Flood.

Why was Puma Punku built? The short answer is no one really knows. However, considering all the findings and discoveries, with all the intricate water channels, the complex stone carvings, and very sophisticated matrix die-cast material, a logical, if not fantastic, conclusion is that it was a giant sleuthing facility to help extract metal from rock. After all, *Titi* in the Aymara language means *metal* (thus Lake Titicaca, "lake of metal"). The ground nearby and in Peru is very rich in copper, tin, gold, lead, and other metals. I suspect the reason for building this site was completely functional, and perhaps thousands of years later, the local Indians made it a religious shrine, for reasons too obvious to mention.

This is very similar to what the Romans did at Balbeck (what is now Lebanon). They found a platform area with monstrously huge carved stone blocks weighing up to 1,500 tons. Most any engineer would acknowledge (as the Romans did) that brute labor stops and superior technology begins at somewhere significantly less than 1,500 tons. The area became a most sacred place in the Roman Empire. The most beautiful Temple of Jupiter was built there. Obviously, acknowledgment, reverence, and respect for such places are not limited to unsophisticated people. (A good reference for further reading is Zecharia Sitchin's book *The Lost Realms*, 1990.)

SACSAYHUAMAN

High up in the Andes mountains, about twelve thousand feet above sea level, beyond the ancient Inca capital city of Cusco, in Peru, South America, lies the ancient fortress known as Sacsayhuaman. From an engineering perspective, the construction is nothing less than stunning. You will find thousands of massive tractite blocks, weighing fifty to one hundred tons, cut or molded perfectly into a jigsaw pattern. The shaping is so perfect that no mortar is required. Neither a sheet of paper nor a razor can get between the stones. If you were to ask a construction engineer to build a walled fortress to withstand things like earthquakes and military attack, the answer would be in two parts: 1) Build a solid-granite walled structure, which would be close to impossible, or 2) build a jigsaw sequence of perfectly molded and cut granite, which is not quite as impossible.

Author at Sacsayhuaman

Whoever designed and built the Sacsayhuaman complex chose the latter (except instead of granite they used tractite, because of its gripping strength). We are told that this complex was built by the Kilcade people in AD 1000. Of course and as usual, how it was constructed is apparently not important to most archeologists, as the Kilcade had no metal to speak of—only gold, and gold cannot cut these stones. If you were to ask the local shamans in the area, they would state very clearly, "We did not build this." Local legend, as told by shamans such as Jorge Louis Delgado, states that this site was built by the "space brothers" thousands of years ago. Jorge Delgado grew up in the Lake Titicaca area (the most sacred area in South America) and was trained by Kallahuaya shamans and Q'ero elders native to the Cusco area of Peru. Sacsayhuaman means *falcon's head*. The legend states that a large falcon had a chemical in its beak that was able to melt stone and thus mold it into place as seen in the jigsaw pattern. In fact, some of the stones do show evidence of exposure to extremely high temperatures, which softened them and allowed them to be molded into position (a technology obviously not available to the local Indians).

Is it possible that the "falcon" was a flying aircraft, capable of individually melting the stones into place? Considering the clear and unequivocal facts on the ground, I do not believe the local legends are to be ignored, as Heinrich Schliemann would testify. After all, he used a myth (the *Iliad* and the *Odyssey*) to discover Troy. Again, now in Peru, we have evidence of a higher technology grossly disproportionate to the social and cultural evolution of the place and period, indicating interference from a vastly superior technology and very advanced species.

However, again we are told to believe they were just lucky. We are told that local Indians, without advanced tools or hard metal, shaped these huge stones and lifted them into place without any evidence of mistakes.

It is also interesting to note that on a plateau near Sacsayhuaman lies a great stone in its rough state. This huge stone is called the *weary stone*. Local Amauta legend, as told by Garcilaso de la Vega (1539–1616), is that local Indians attempted to move this stone many years ago just to see if they could do it. Twenty thousand men were said to have been used to move it, and during the process, three thousand died. It is for that reason the stone is where it is. You can draw your own conclusions.

One needs to consider that fact that these sacred places are sacred because of what was found there and that this is evidence of a superior culture and advanced technology long lost to the historical record. One thinks of Baalbek in the highlands of Lebanon, which became the most sacred place in the ancient Roman world. The great Temple of Jupiter was built there. Why was it so sacred? Did some priest simply say it was, for no particular reason? Of course not. It is because at Baalbek Lebanon, you will find a huge complex of granite blocks weighing as much as 1,500 tons. That is why it is sacred. It stares you in the face and says *figure me out if you think you're so smart*. The stones are shouting at us. Listen to them. The Romans knew that 1,500 tons of pure granite perfectly placed into position goes far beyond human strength alone and strongly indicates a highly sophisticated ancient technology, heretofore unknown.

Author's photo

TEMPLE OF THE MOON/CAVE OF THE SERPENT

Sacred Because of Functionality

It may be interesting to note that a ten- to fifteen-minute drive from Sacsayhuaman is the Temple of the Moon (popularly known as the Cave of the Serpent). A shaman there told me that this is the original temple of the moon and that all other namesake temples are based on this one. He told me that this cave/temple was constructed by the gods long ago to help humans survive. It is not one of the typical tourist sites, so my friend Gary Rosenthal and I hired a taxi to take us there. We were alone except for a local shaman (who, by the way, refused to take any money for his excellent instruction and obviously was a very holy man). The site is obviously very old, but the exact date is unknown.

This temple is actually a cave, and deep inside there is a large stone table that was polished smoothly. The table is slightly tilted toward the opening of the cave and is part of the granite structure of the cave; it cannot be removed nor its position changed. Above the table is an opening to the sky. We were told that during the Peruvian winter solstice (June 22), the moon shines through the ceiling hole onto the table and illuminates the entire cave, so much so it can be seen from the outside entrance.

What is very peculiar about this temple is the fact that, clearly visible to the right of the front entrance, there is an ancient carving of an elephant head. Academic archeologists think that the site is only about one thousand years old. But paleontologists and anthropologists almost universally agree that elephants, mastodons, and similar mammals became extinct in the Americas about ten thousand years ago, with the ending of the last ice age. So either this cave was made, like, *yesterday*, or it is very, *very* old indeed.

To modern man, knowing the day of the winter solstice is irrelevant. But for primitive man, it was life and death. One needs to know when to plant, and knowing the exact day of the winter solstice can be very helpful. Thus, having a permanent fixture to shed light on the problem (no pun intended) can be extremely helpful.

In addition, present-day Peruvian shamans freely admit that their ancient relatives had help from what they call "space brothers." They also teach that their most sacred god, Veracocha, taught them a more peaceful existence, which included terrace farming and animal husbandry.

Is this not evidence of "divine" assistance? Is this not evidence that we were never alone?

THE GREAT PYRAMID OF GIZA

Much has been written about the enigmas of this most fantastic work of engineering, and I don't feel I have much to add. But no discussion of archeological enigmas is complete without some discussion of the Great Pyramid of Egypt. Therefore, I will share some salient examples of what we know and don't know.

The Great Pyramid is colossal. Measuring about 755 feet at its base and approximately 481 feet high, its faces are directed perfectly east, west, north, and south. It sits perfectly on the thirtieth parallel (29 degrees, 58 minutes, 51 seconds, an error of less than an arc minute; when considering atmospheric refractive error, this may not be an error at all). This was all apparently done without sophisticated instruments. Perhaps they were just lucky. No reasonable explanation has ever been given for how this amazing feat was accomplished, but nonetheless, it is there for all to witness, as if challenging us: "Can you figure me out if you're so smart?"

It is composed of approximately 2.3 million blocks of stone weighing an average of 2.5 tons each, indicating that the entire structure weighs an astounding 6 million tons. Most of the larger blocks are at the base, but in apparent defiance of engineering practice, some larger blocks, weighing ten to fifteen tons each, are placed high up at the thirty-fifth course.

The Great Pyramid is amazingly stable, while all others, newer and smaller, are falling apart.

High up and deep inside the pyramid is the King's Chamber, with about one hundred blocks lining the walls, each weighing seventy tons. The stones of the ceiling weigh fifty tons each. These blocks are granite (from a quarry five hundred miles away), which is much harder than the limestone encasing the pyramid. Examination of this granite reveals they are polished to the accuracy of a human hair. This was obviously not done with brute force alone.

But this is not all.

In the late nineteenth century, the famous British Egyptologist Sir William M. Flinders Petrie carried out a detailed study of the artifacts from the Great Pyramid and concluded that it was constructed with much more modern technology. Power tools. The evidence is there for all to see at the Petrie Museum of Egyptian Archaeology, University College, London. In that museum, you will bear witness to the famous granite "Core #7," which was clearly and unequivocally made with a power drill. These observations of Petrie were validated in 1999 by Christopher Dunn, an engineer and master craftsman in the aerospace industry. He discovered the pitch to be 0.11 to 0.12 inches and the groove to be between 0 and 0.005 inches deep. This is just one example of many.

In 1988, Christopher Dunn examined the inner surface of the granite coffer in the King's Chamber of the Great Pyramid with a *parallel* (a machinist device that determines the smoothness of a surface). The flatness of this surface was within one ten-thousandth of an inch throughout the inner surface and in all directions. That is less than a human hair, and don't forget—this is the *inner surface*, which is much more difficult to do technically. Clearly, this is where brute force stops and technology begins. And these are not the findings of fringe individuals with wild speculations and swooping generalizations. These are nuts-and-bolts engineers and scientists with no political, religious, or anthropological axe to grind. In the words of Thomas Jefferson, they are just "following the science wherever it may lead."

Anyone will testify that the Great Pyramid is beautiful to look at, not only because of its size but also because of an inherent perfection in its construction. Examples of the mathematical perfection of the Great Pyramid are almost endless. If the height of the pyramid were to be taken as the radius of a circle, the circumference of the circle would perfectly enclose the perimeter of the base. This is known as "circling the square," and doing so strongly indicates that the designer had knowledge of pi (the ratio of the circumference of a circle to its diameter), something the ancient Egyptians (or anybody else at that time) were not supposed to know. In order to achieve this, the angles of each of the sides must be *exactly* 51 degrees, 51 minutes, and 14.3 seconds.

The Great Pyramid site is the exact center of the Earth's landmass. The east-west parallel and the north-south meridian crossing at that site divide the planet into equal quadrants.

The southern shaft, coming from the King's Chamber upward to the side of the pyramid, is exactly 45 degrees (0 minutes and 0 seconds) and points exactly to the Al Nitak star on Orion's belt (as it appeared in the sky circa 2450 BC). This star was believed to be the original home of the Egyptian god Osirus.

No one was ever buried in the Great Pyramid. There is no writing on the walls, nor are there the paintings typically found at other Egyptian burial sites. We really do not know who built it or who designed and supervised its construction. It is just there and by its physical appearance makes a bold statement to anyone who would listen: *Try to figure this enigma out.*

Yet we are led to believe that primitive man walked out of the Neolithic period, without a wheel, without a pulley, without knowledge of either mathematics or geometry, without sophisticated instruments or metal alloys, and without engineering skills or sophisticated tools; fashioned some of the hardest substances on Earth to an accuracy not seen until the advent of modern power tools; and constructed the largest, strongest, most durable, and most perfectly designed monument in the world. And it took them only twenty years.

I don't think so. They were not just *lucky.* To me, it is clear that the statement of the Great Pyramid is this: *You are not and have never been alone.*

Artistic representation

THE STATUE OF KING GUDEA
And the Universal Measure

In the British Museum in London and at the Louvre in Paris, you will find a statue of an ancient Sumerian king called Gudea. It was discovered in Mesopotamia by Ernest de Sarzec in 1880. It's about four thousand years old. It is beautiful, and the workmanship is spectacular.

The one in Paris is of the best quality, as it is complete. The statue in London is missing the head.

It is made of diorite. Diorite is extremely hard. In fact, the only thing harder than diorite is diamond. No one really knows how it was made, but nonetheless it is there for all to see.

On the side of the statue is a carefully graduated measurement for a half *kush* (a sacred measurement for the Sumerian people). It is gauged to be about 24.97 cm. It is also called a *barley cubit*, which equals 180 barley seeds lined up belly to belly. The barley seed is very interesting, because barley DNA is very stable—it hasn't changed much with time. The measurement was handed down from the ancient kings and, presumably, from the gods. Anyway, the sacred measurements for the ancient Sumerians were the half kush, kush, and double kush.

In the late 1700s, during the period known as the Age of Enlightenment, the French decided to create a unit of measurement that would be unique to this Earth, and the formula for this measurement would be so logical that it could be refined as technology improved. This idea was derived from the ancient Greeks, who referred to this concept as a "universal measure." From this came the term *meter*.

Many great minds were solicited, including Thomas Jefferson. He believed that the best method was to amalgamate time and distance unique to this Earth. Knowing that there are sixty seconds to a minute, and sixty minutes to an hour, and twenty-four hours to the day (of course, another gift from the ancient Sumerians), one concludes that there are 86,400 seconds to the day. Knowing that the beat of a pendulum is constant depending on the pendulum length (remember the grandfather clock), then why not have a pendulum machine at a certain location on Earth (preferably at the equator and at a certain longitude) that would be connected to a mechanical counter? Thus, as the sun rises, with the assistance of a sextant (a device used for marine navigation), the pendulum length can be adjusted from day to day, until the mechanical counter equals exactly 86,400 seconds from sunrise to sunrise.

Thomas Jefferson was not the first to come up with this idea. Christopher Wren also suggested a similar approach in 1668. Almost all agreed that this approach was logical but also technically and geographically difficult (which of course means *expensive*). Therefore, it was decided that this universal measure would be one ten-millionth the length from the North Pole to the equator, as this could be measured entirely in Europe.

So this became the definition for the *meter*—until 1983, when an atomic definition was developed: the distance traveled by light in a vacuum in $\frac{1}{299,792,458}$ of a second. Clearly, it was a mathematical formula to fit the current accepted length of a meter and not the more pure and idealistic version suggested by Jefferson.

HOW IT WORKS

There is a formula for determining the length of a pendulum and the time of a pendulum period (a complete cycle, or two sweeps) or beat (a half cycle, or one sweep).

$$T = 2\pi \sqrt{\frac{L}{g}}$$

L = length of the pendulum
G = Earth's gravitational pull

If you make L equal to the half kush (24.97 cm), then the time of one pendulum period (two sweeps) is 1.003 seconds. If you make L equal to a double kush (99.88 cm), the beat (one sweep) is 1.003 seconds. It is obvious that our ancestors had significant help, and through that help, the helpful ones (ancient aliens) were sending a message to anyone who had the scientific and technical expertise to listen.

But this is not all.

The Earth moves exactly at sixty thousand kushes per second around the sun. This is one ten-thousandth the speed of light. (The speed of light, by the way, is an even 600 million kushes per second.)

Sexidecimal—decimal—sexadecimal. The way of counting by the ancient Sumerians. (1 to 10, then 10 to 60, then 60 to 360, then 360 to 3600.) Absolutely spectacular. It could not have been more perfect. It is nothing less than astounding to conclude that while mankind, on the dawn of the industrial age and the modern era, was looking for the perfect measuring stick that would amalgamate time and gravity unique to this planet, it had already been given to us by the "gods" thousands of years before the birth of Christ.

These are not random numbers, formulated by ignorant farmers who just walked out of the Megalithic period. This is not the result of luck or serendipity. There is clearly a message here, provided we are intelligent enough to appreciate it.

Yes, my friend, we have been, and continue to be, visited by a very sophisticated class of beings. Indeed, they have never left. They have helped us in our survival and have given us clues to their existence for future generations. It is as if the very stones are crying out, telling us to examine them, learn from them, and cast aside previous narrow-minded, ill-conceived beliefs.

A very good reference for further reading is *Civilization One: The World Is Not As You Thought It Was*, by Christopher Knight and Alan Butler (2010).

Conclusion

Again, we have evidence of assistance from advanced beings, either helping humans develop incredible structures or developing these structures themselves, without humans, evidence that man was part of a sophisticated culture much older than previously thought. This is not a fringe perspective. It is a *finding*, based on evidence— evidence suggesting ancient aliens developed incredible things.

Why? For posterity? Are they trying to tell us something? Are the stones crying out? I suspect they are.

Chapter Six

The UFO Phenomenon
The Gods Have Never Left

Space and time are principles that we live by; they are not fundamental laws of the universe.
–Albert Einstein

Wisdom is found only in truth.
–Johan Wolfgang von Goethe

The size and age of the universe encourages the belief that many technologically advanced civilizations should be plentiful if not ubiquitous. However, this belief appears to be inconsistent with the observational facts. Therefore, either 1) the initial assumption is incorrect, 2) our current observation is incomplete, or 3) our search methodologies are incorrect.
–The Fermi Paradox (named after Enrico Fermi, nuclear physicist)

PROPER HUNTING REQUIRES PROPER BAIT

As a child, I recall visits with my father to Veterans of Foreign Wars gatherings. There I would hear World War II stories from the army as well as the navy perspective. For some reason, I was very interested in the nautical history of the war. Often there was a great deal of discontent with Admiral Ernest J. King and his initial handling of the U-boat menace on the eastern shore of the United States.

Admiral King was a fleet commander and chief of naval operations, as well as a member of the Joint Chiefs of Staff. In other words, he ran the navy during World War II. When the United States became involved in the early months of the war, the German U-boats had a field day off our eastern shores, from Florida to New York.

Great Britain had much more experience with German submarines than Admiral King did. The English had learned the value and safety of sailing their cargo ships in a convoy of multiple ships, with escort protection from destroyers. They also knew about blacking out their coastline at night so submarine commanders could not use the coastline to silhouette their targets.

But Admiral King ignored English offers of advice. The reason for this (it is believed) is that Admiral King was an Anglophobe. He just did not like the English.

So he sent his destroyers out helter-skelter looking for U-boats. If he ever used a schedule, the U-boat commanders soon learned it and easily avoided our warships. In fact, in the first two months of 1942 alone, over 2 million tons of US shipping—and not a single U-boat—were lost. Hundreds of sailors died as well, making this the single greatest maritime disaster in US history.

It was not until King appreciated the concept of a convoy that he gained any success in finding and sinking U-boats. By the end of the summer of '42, things became very different for the German submarines. Thus, this late-summer change in US nautical strategy became known to U-boat commanders as the "end of the happy times."

Why is this story important? In his bestselling book *Universe, Life, Intelligence* (1962), Carl Sagan proposed that the best way to prove intelligent life in the universe is to build powerful radio telescopes that may pick up signals of communication in the microwave spectrum emanating from outer space. So the Ohio State University, with a $71,000 grant from the National Science Foundation, built a huge radio telescope called Big Ear in 1963, and the SETI (Search for Extraterrestrial Intelligence) program began. It still runs today but is no longer funded by the US government.

Of course, they never really found anything. This is simply because, I believe, the *search methodology* is wrong. You might say they were "looking for love in all the wrong places." Like Admiral King in World War II, they went looking helter-skelter—fishing without bait or a lure. King learned that the convoy was the bait. And where you have bait, you're likely to have fish.

I have always believed that this big beautiful planet, with its abundance of biological activity, mineral resources, and beautiful mountains and shorelines, is the proper bait and is unlikely to be missed by extraterrestrials. If you want to find extraterrestrials, look to Earth. That is exactly what the UFO hunters are doing today, and I have little doubt that they will get there first.

So, assuming that extraterrestrials are already here, what are they most concerned about, and where can they be found? Well, if you discover the former, you will most likely arrive at the latter. By understanding their "bait," we can draw conclusions regarding their concerns and motives.

Let us accept some logical sequential postulates.

(1) Extraterrestrials are here and have been here for a very long time.

This logically leads to postulate two:

(2) They are highly advanced technologically, and there is very little that we can do for them,

other than perhaps serve as servants or slaves. (Hopefully, they are vegetarians.)

Given postulate two, and the fact that in interspecies relations "might makes right," we arrive at postulate three:

(3) They have the right to claim this planet as their own.

Which leads to postulate four: Because ownership fosters material concern and responsibility...

(4) They have legitimate concerns about the biological viability and health of this beautiful and hospitable planet.

If they have an interest and can make a claim but haven't done so yet, then:

(5) They are under orders not to make themselves known. Otherwise, they would exploit us (as we would probably do to them if we had the upper hand).

Postulate four is the most relevant. What would be their primary concern about the biological viability and health of this planet? The single and most obvious conclusion is the advent of nuclear weapons. A major nuclear event can turn this planet into a lifeless wasteland unfit for habitation, extraterrestrial or otherwise. We have the potential to destroy this planet many times over, and that would, logically, be of tremendous concern to them.

UFOs

Of all the topics related to the golden thread, none is more filled with deception than the unidentified flying object (UFO) phenomenon. Simply witness the photos and videos on YouTube, of which almost all are obvious hoaxes. These, however, do not disqualify the existence of UFOs. There are many credible testimonies to this phenomenon, as well as some physical evidence.

The evidence for the existence of UFOs as an ultra-technological and thus extraearthly phenomenon can be organized in three categories.

1. A photo or video that holds up to technical and scientific scrutiny. This is very rare, and even if there is no obvious technical or scientific fault, photo or video evidence alone is always suspect.

2. Observation by a credible witness whose integrity, technical knowledge, and experience is unquestionable.

3. Multiple witnesses, radar confirmation, physical evidence, or other credible corroborating information.

To wade through the mire of thousands of hoaxes in the UFO literature can be so boring that it's often only good for a laugh or to treat insomnia. How can one determine the relevant from the irrelevant cases?

If you question this, check the website for the National Investigations Committee on Aerial Phenomenon (NICAD). There you will find the numerous accounts from military and civilian personnel that will convince you that these are not isolated events but a series of visitations and interventions by UFOs.

I have reviewed thousands of cases, and here I present some that I believe are relevant and even compelling. However, first I would like to report a more personal experience.

THE DAY THE WORLD ALMOST BLEW UP

I am reporting this for the first time on paper.

In the 1980s, as a young neurologist, I had the opportunity to take care of many famous persons. During this time, I became the primary neurologist to the director of the CIA, William Casey.

William Casey was the most interesting man in the Reagan administration. He was a super-spy in the early days with "Wild Bill" Donovan in World War II. His field of interest was dropping spies into Germany. I would say he was the most intelligent man (among many) of that administration. If he gave you a book to read, regardless of thickness, he would expect you to have read it, in its entirety, by the next morning.

Sadly, when I met him, Mr. Casey was dying of a large cerebral lymphoma (brain cancer) in the left hemisphere. He underwent surgery, but this resulted in weakness on the right side and difficulty with speech (expressive aphasia). He knew all too well he was dying and took a liking to me, this young neurologist. We had many late-night discussions about foreign policy and international intrigue. In fact, I was at the meeting that decided Robert Gates was to replace him as the next director of the CIA. I was even invited to late-night visits to the Oval Office on occasion.

I confirmed with Dr. Thorp, the medical director of the CIA at that time, that William Casey was not under duress, nor had his arm been twisted in any way. For reasons not completely known to me, he requested that I be indoctrinated into some of the more secret historical facts of the CIA. The most intriguing was Project Jennifer.

If you are old enough, you might remember a Soviet submarine that was "accidentally" sunk in the Pacific Ocean. The submarine was a Golf-class diesel electric, a near copycat of the type 21 German U-boat of late World War II, and it sunk on March 7, 1968, a few hundred miles from Hawaii. Although the submarine was old technology, it did carry three Soviet R-21 nuclear missiles, located just aft of the conning tower. These missiles had a trajectory in excess of three hundred miles.

In those days, the Soviet system had a complicated chain of command. Although the belief at the time was that Leonid Brezhnev was chief of state (general secretary of the Communist Party), he did not have all the power. The real power was in the hands of a man named Mikhail Suslov. If Mr. Brezhnev was a communist, Suslov was a communist to the extreme. He believed firmly in the eventual domination of the world by the Soviet system. He was a kingmaker and a king breaker. He was ruthless and calculating. Known both as the Red Professor and the Grand Inquisitor, Suslov, with henchman Uri Andropov at his side, was responsible for many purges and deportations.

Suslov had a devious plan, and in order for it to come to fruition, he placed Uri Andropov as head of the KGB, in early 1967. Although the CIA file listed him as of Jewish extraction, he was an atheist and an adamant communist. He was also very intelligent and, like Suslov, was an ideologue believing in the eventual Soviet takeover of the world. Head of state Brezhnev, although known to drink hard and get loud at times, did not share this risky view, and this was a point of contention between him and these other two.

There was one serious problem for Brezhnev at that time. Unlike in the United States, all Soviet nuclear weapons were under the control of the KGB.

In late 1966 and early 1967, Uri Andropov and Suslov decided on a bold plan of action that would assure Soviet world domination. They had been sending nuclear bomb material and Golf-class submarines to China. If they could park a Soviet submarine close to Hawaii, fire a missile on the American islands, and have enough hard evidence implicating the Chinese, then the United States and China would have it out, and the Soviet Union would inherit whatever was left in the aftermath.

Sounds simple enough, but Suslov did not know there was a fail-safe mechanism installed (with some help and encouragement from the United States) on Soviet submarine missiles, which would blow the top portion of any missile that was fired without the final code.

This is precisely what happened. The rear conning tower exploded, and the sub K-129 sank in thousands of feet of water.

To make a long story short, the sunken sub was found with the help of the *USS Halibut*, an old nuclear sub that was converted into a spy sub. Thousands of underwater photos were taken, which elevated the index of suspicion that an attack was attempted. The navy representative who found the sub wanted to pursue further, but President Nixon believed that the CIA was better at keeping secrets. Billionaire Howard Hughes was contacted to build the *Glomar Explorer* (a huge mining ship with a large moon pool and very tall derrick) to excavate the wreckage, and thus Project Jennifer was started. President Ford kept it going after Nixon left office.

Contrary to popular misconception, they really did get pretty much the whole boat. The sub was disassembled, taken to a large hangar in San Francisco, and extensively photographed. A diary from a Soviet sailor was also retrieved. (Navy sailors are told not to produce diaries. Great blackmail material if ever there was.) In summary, we learned in March of 1968 that we almost had a worldwide nuclear holocaust orchestrated by Soviet communist ideology.

It is very clear to me now that aliens (Anunnaki) knew of this plot and needed to see, should something like this happen again, whether they really could shut down our Minuteman intercontinental missiles on a moment's notice. So they sent a UFO to Oscar Flight and Echo Flight in Montana (among other sites) and found out they could. They are obviously very interested that we don't destroy what they consider *their* planet with a nuclear holocaust.

(The incident with K-129 also explains why President Nixon, the vehement anticommunist, did something very uncharacteristic: he visited Mao Zedong, opening doors to the West. Because of the high risk of a global nuclear holocaust, and seeing firsthand the danger of a lack of communication, Nixon buried the hatchet and opened up China for commerce, and I believe we are all a little safer for it.)

These reports are also relevant because military personnel are actually reluctant to report UFO sightings. Such reports are strongly discouraged from military command and almost guarantee a dismissal or a stop to one's career. This is due to the Personnel Reliability Program (PRP), in which UFO reporters are required to have a psychological examination, which often results in a blip on their record or even dismissal. Obviously, UFO witnesses are very intimidated by their superiors. This has been one of most important tools used by our government to keep the UFO phenomenon under wraps. (I suspect the government does this because it is responsible for nuclear security, and any evidence suggesting that it does not have complete control poorly reflects on its competence or may result in national panic. I believe today the average citizen is ready for the truth. Nonetheless, for reasons of security or otherwise, our government has been lying to us.)

THE UFO INCIDENT AT THE 490TH STRATEGIC MISSILE SQUAD MARCH 16, 1967
The UFO ICBM Connection

This incident clearly fits into the third category of UFO evidence. Testimony of the event is given by Captain Robert Salas, a highly educated and experienced officer in the air force and graduate of the Air Force Academy. He served active duty from 1964 to 1971 and worked as an engineer on the Titan 3 missile and for twenty-one years for the FAA. His testimony is all over the Internet, and he, in collaboration with Jim Klotz, has written a book, *Faded Giant*, on this topic.

Robert Salas states that on the morning of March 16, 1967, he was a base commander on duty at the 490th Missile Squadron at Oscar Flight, one of five launch control facilities for the Minuteman 1 intercontinental nuclear missile. He was sixty feet below ground level at the launch control facility when he received a call from the topside guard, who had observed some strange lights moving about the launch facility. Nothing much was made of this report, and the conversation was closed.

Then, about thirty minutes later, he received another call, this time from a much more frantic and obviously frightened guard. He reported that a large, noiseless, glowing red object was hovering over the front gate, and all the guards had their weapons drawn and pointing at the craft. At that same time, all the missiles at the facility (there were about eight or ten) one by one went into a no-go condition, with lights flashing and bells ringing, meaning that they could not be fired. This had never occurred before, because all missiles had their own independent launch control mechanisms, including battery backup. A shutdown is obviously very serious, and for the technicians to get them back up and running (alert status) takes all day.

Salas called topside again, and the guard told him that the object flew away at high speed. The commander then called the command post and was informed that the same thing happened to Echo Flight, a similar base fifty or so miles away; they had lost all ten of their missiles. There is also some evidence that other launch facilities were affected in the same manner on that date.

Robert Salas was joined by many credible officers and airmen in witnessing this event. These men were handpicked for their integrity, competence, and trust-worthiness. I doubt you could get more credible witness-es for anything. These men know the difference between a helicopter and an airplane. They know when some-thing is not right and are very unlikely to create a hoax.

There is also collaborating witness testimony from First Lieutenant Robert C. Jamison (combat targeting team commander), of 341 Missile Base Maintenance at Malmstrom AFB, Montana. He has testified that he assisted in the restart operations of these missiles at Oscar Flight after they were made no-go by the UFOs. He was instructed that, before going out to examine the shut-down missiles, he was to remain at the Maelstrom base, until the UFO sightings had ceased, for safety reasons.

It is interesting to note that UFO sightings at nuclear intercontinental missile launch control facilities are *not* a rare phenomenon. There are now well over one hundred testimonies from US Air Force personnel (some highly decorated and with advanced ratings) regarding UFO sightings and occasional interference at US nuclear missile facilities. If you were to review Robert L. Hastings's extensive research on the matter, you would find that UFO sightings at nuclear missile launch facilities are a fairly common phenomenon. In fact, if you really wanted to see a UFO, it would seem that this is the place to go.

The Salas testimony is available on YouTube, and I recommend that you check it out. He said that he is willing to take an oath before Congress and that he and many others in the know are willing to testify (over one hundred). Obviously he is not lying.

The Salas testimony is very remarkable because it involves a direct intervention from an unidentified flying object, not just a sighting. This UFO presented itself to the launch facility, shut down the facility's missiles, and then flew away at an extremely high speed. All the UFO sightings before and after March 1967 were just sightings and nothing more (as far as we know).

There is the possible exception of what happened at Ellsworth Air Force Base in South Dakota in July 1966. There was a power supply failure as well as power backup failure, in which a Minuteman missile went off alert status at launch facility Juliet 5. A UFO was observed resting on its tripod landing gear inside the security gate and emitting a bright light that illuminated the entire area. Whether the UFO caused the power failure is not completely understood. As a helicopter approached the site, the UFO lifted off the ground, flew away quickly, and disappeared.

The incident at Oscar Flight, then, is still the only confirmed case of intervention with a military installation. And the date, March 1967, is very relevant.

OPERATION MAINBRACE AND THE *USS ROOSEVELT*

UFO sightings are not limited to land. In support of our theory—that these extraterrestrials are deeply concerned about the potential devastation of Earth's biosphere by the irresponsible deployment of nuclear weapons—there is another event of compelling interest.

The *USS Franklin D. Roosevelt* was the second of three Midway-class aircraft carriers. She was commissioned on April 29, 1945 (shortly before the end of World War II), as the *Coral Sea*, but after the death of President Franklin D. Roosevelt, President Harry S. Truman honored him by changing the name to *Franklin D. Roosevelt*. In 1946, she became the first aircraft carrier to carry a full complement of jet aircraft. Then in 1950, she became the first US aircraft carrier to carry nuclear weapons, and later thermonuclear weapons (hydrogen bombs), to sea—something of definite interest to our extraterrestrial friends.

Things were very tense at that time. China became communist in 1949 and developed a nuclear device of its own. Russia (also with nuclear weapons) was flexing its muscle and securing its Eastern Bloc puppet regimes, creating deep resentment among her former allies.

With all this tension going on, NATO, in September 1952, began its first massive naval exercise in the North Atlantic, known as Operation Mainbrace. The *USS Roosevelt* and its full complement of jets and nuclear weapons participated, along with three other US aircraft carriers, two major US battleships, and all the naval vessels the other NATO countries could muster. They headed for the North Atlantic under the direction of Sir Patrick Brind, a British admiral.

On September 13, Lieutenant Commander Schmidt Jensen and some members of his crew reported a triangular craft buzzing about his ship, emitting a bluish glow, and traveling at an estimated speed of about nine hundred miles per hour. (Remember, this was before the supersonic age, and no jet craft could exceed much more than five hundred miles per hour.) During the next week, there were at least four more UFO sightings.

One sighting included the crew of the *USS Roosevelt*. They described a silvery spherical object moving across the sky at a high rate of speed. Photos were reportedly taken and sent to naval intelligence. These photos were never made public. According to Air Project Chief Captain Ruppelt, the photos were "excellent."

On September 20, three Danish air force officers at Karup Airfield in Denmark reported a UFO, described as a shiny metallic disc, flying right above them at a high rate of speed in the direction of the fleet. Other UFOs were seen over Sweden and Hamburg, Germany, during that same period. It should be noted that these UFO reports were made by competent and experienced naval and air force personnel well acquainted with seeing objects in the sky. They would know the difference between a balloon and a fighter jet and would be expected to know if something was not quite right.

The *USS Roosevelt* had many UFO sightings later in her lifetime. One interesting sighting was reported by Chester Grusinski, a crew member aboard the *Roosevelt* in 1958 near Cuba. He described the UFO as spherical (or tubular) and about seventy-five to one hundred feet long. It was an orange-red color, and he could feel the heat on his face. He could also see silhouettes of figures inside the craft, looking out. He said they were not human. The bottom turned "cherry red," and it was gone in a "flash."

UFOS IN RUSSIA

It should also be noted that UFO incidents are not limited to the United States. The former Soviet Union has also experienced many UFO interactions and has had a particular interest in the unidentified submersible object (USO) phenomenon as well. Paul Stonehill has been researching recently declassified Soviet military files and discovered that they were equally, if not more, interested in UFO sightings as counterparts in the United States. He has published a book, *Soviet UFO Files*, and is working on another as he continues his important research.

One of these files reported an attempt by the Soviet military to capture a USO. Unfortunately, men died in the attempt, and the order was given not to do it again.

It is interesting to note that these vehicles and their pilots were referred to by the Soviet authorities as "watchers/observers." This sounds like something from the ancient book of Enoch, which I doubt very much they have read. The Sumerian word *IGI* also means *watchers*, which is the same as the Egyptian word for *gods* (NTR). Nothing is new under the sun.

Does this consistent naming suggest a basic sub-conscious memory from the human experiences of the distant past? I suspect it does, and one must admit that it is uncanny.

FRANCE (THE COMITA REPORT)

One cannot discuss the international UFO phenomenon without mentioning the French COMITA report. This was a study performed by a nonprofit organization. COMITA in French stands for *Committee for In-Depth Studies*. The report was unsolicited by the French government, but no one can question the high-level officials who are signatories to the document, including scientific, engineering, political, governmental, and military luminaries, such as:

General Bruno Lemoine (French air force)

Admiral Marc Merlo

Denis Blancher (chief of the national police and superintendent of the Ministry of the Interior)

General Alain Orszag (PhD physicist)

Christian Marchal (chief engineer and research director of the National Office of Aeronautical Research)

The idea for this UFO investigation started in 1995, when General Letty approached General Norlan, at that time director of IHEDN (Institute for Advanced Studies for National Defense), was approached by General Letty. Letty wanted to know whether UFO phenomenon posed any threat to French national defense.

There would be full cooperation. The research included papers and films from GEPAN_SEPRA, the official government agency for UFO investigations. The comparison to the United States would be as if the Department of Defense, NASA, and the CIA asked the Heritage Foundation to study the UFO phenomenon and promised to give full support with all its documentation.

In 1999, the ninety-page COMITA report was first sent to French President Jacques Chirac and to Prime Minister Lionel Jospin. Of course, when it went public, it went down like a lead balloon. Many politicians headed for the hills to cover their asses, saying, "I didn't ask for this!" Well, I'm sorry. Whether they asked for it or not is immaterial. It is there now for all to read.

The gist of the report is that 5 percent of the UFO cases they studied were absolutely unexplainable. The best theory was that they were indeed of extraterrestrial origin, and the US government was engaged in a massive UFO cover-up.

THE RENDLESHAM FOREST INCIDENT

Rendlesham Forest is an area between the twin (US Air Force and RAF) nuclear weapons storage facility in Suffolk County, England. At the time of this incident, it was the largest nuclear weapons storage facility in Europe.

Confirming testimony is given by Charles I. Halt, lieutenant colonel and deputy base commander. His original testimony of January 13, 1981, has been declassified and is part of the public record.

Halt reports that in the early morning hours of December 26, 1980, a two-man security patrol near the east gate of the RAF facility observed strange lights above the forest. They phoned in the report, assistance was granted, and an extra security detail was sent out. It was first thought that an aircraft was down and on fire. Three men of the security team (Jim Penniston, Ed Cabansag, and John Burroughs) drove as close to the sight as possible and then went on foot on the frozen terrain. They then realized that it was not a burning airplane but rather a large object brightly lit.

Security Officer Penniston reportedly went the closest to the object. He described the object as triangular and pyramid shaped, measuring about nine feet long. It had a dark black glass-like texture and triangular landing gear. There was a bluish-white glow to the craft and a blinking red light on top. He noted some hieroglyphic markings on the surface, a seemingly symbolic drawing of a pyramid circumscribed by a circle, with two solid dark circles—one on the lower right of the pyramid, and the other, slightly smaller, on the upper left part of the circle. He later presented drawings of his findings.

He reports that he actually touched the craft, and he notes that there was a strange sensation that his "movements did not feel normal" when near it. He and the others also reported that local farm animals were making a lot of noise.

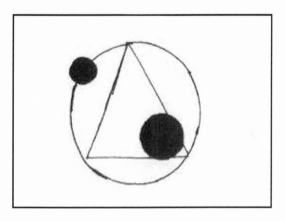

Artistic representation

Two nights later, in the early morning of December 28, the strange lights were seen again. Five men investigated, this time with Deputy Base Commander Halt. He brought a camera, a tape recorder, and a powerful mobile generator to light up the area. Halt's plan at the time was to dispel any thought that this might be a UFO.

Handheld communication devices were fraught with interference, and the powerful lights began to fail. Nonetheless, they did spot a large triangular metallic craft. Some said it rested on a tripod landing gear.

Ground impressions were later found at the site. A Geiger counter reading revealed a radiation level seven times what would be expected. This was later confirmed by Nick Pope, an investigator with the Defense Radiological Protection Service.

Holt's camera film was confiscated, but his tape recording is now part of the public record and can be downloaded from multiple sites on the Internet.

The USAF never took this episode very seriously (or so we are led to believe), but the event was clearly a life-altering one for all witnesses involved. Government critics claim these men simply saw the lights from the Orford Ness Lighthouse, five miles away and in the line of sight from the witnesses' perspective. This, I think, is a little disingenuous. Although the lighthouse was within their line of sight, it is extremely doubtful that it could be mistaken by any professional in the military for a triangular metallic craft resting on a tripod, and if it can be, we need to review the USAF recruiting methodology. There were some minor temporal and spatial discrepancies in the witness accounts, as might be expected with multiple witnesses, but there is no doubt in their minds as to what they witnessed on those cold December nights in 1980.

Lieutenant Colonel Charles I. Halt later testified, "I believe that the objects that I saw at close quarters were extraterrestrial in origin and that the security services of both the United States and the United Kingdom have attempted—both then and now—to subvert the significance of what occurred at Rendlesham Forest and RAF Bentwaters by the use of well-practiced methods of disinformation."

If the testimony of credible professionals it not enough, if multiple eyewitnesses and radar confirmation is what you seek, then let's discuss the Belgium Wave.

THE BELGIUM UFO WAVE

From November 29, 1989, to April 1990, there were almost daily UFO sightings throughout Belgium. UFOs were tracked on radar, photographed, and witnessed by as many as thirteen thousand people. There were over 2,600 written reports. The craft was triangular in shape, equilateral, and with lights at each corner changing in color from red to green to yellow. The lights were brighter than the stars. The craft made no noise.

The sightings were confirmed by local authorities. Major General Wilfried de Brouwer authorized two F-16s to fly to Thorembasis-Gembloux, near Brussels. The F-16s were never able to get visual confirmation but did receive radar confirmation. There was also secondary radar confirmation at Semmerzake radar tower. There were three separate times that the F-16s were able to lock the craft onto their radar, and each time the UFO instantly accelerated to extremely high speed, unlocking the radar. Radar data confirmed the craft once descended from 3,350 meters to almost ground level in about two seconds; later it descended nine hundred meters in less than two seconds. Both changes in speed are enough to kill any earthly pilot.

It was thought that a nearby US Air Force base might be flying a stealth bomber, but a simple phone call made by Major General de Brouwer put that to rest. No stealth airplanes were flying.

Some, such as Renaud Leclet, have suggested the ridiculous idea that these sightings were mistaken helicopters. Noiseless triangular helicopters able to accelerate nine hundred meters in less than two seconds? Able to exceed the speed of sound without a sonic boom? That is even more fantastic than a UFO.

This begs the question: Why Belgium? Belgium is the host of NATO headquarters, and, obviously, NATO has a lot of nuclear weapons. In addition, the Soviet Union, with its thousands of megatons of nuclear devices, was falling apart from the inside. The ruble was worthless. Estonia and Lithuania declared independence. An internal war was brewing between reformers like Boris Yeltsin and communist hard-liners.

This was a period of grave concern. A message had to be sent to the capital of NATO—Belgium—so cooler heads would prevail during this transition, so people would think twice before any knee-jerk decision was made that could turn a revolution into a worldwide nuclear holocaust.

The Belgium Wave and other sightings are excellently documented in Leslie Kean's bestselling book, *UFOs* (2010). This is a must-read for anyone interested in this subject.

THE IRANIAN UFO INCIDENT

This remarkable incident was reported by General Parvis Jafari (retired) of the Iranian air force. As the report goes, a bright object with changing colors was seen hovering about six thousand feet above the city of Tehran in September 1976. A phantom F-4 II jet was scrambled from Shahorkhi Air Force Base. The jet had two pilots seated in tandem, Captain Aziz Khani and navigator First Lieutenant Hossein Shokri. The onboard radar estimated the size of the object to be about the size of a Boeing 707 tanker.

Then the captain noted a smaller object coming out of the larger one and heading straight for the F-4. The pilot admitted that he became terrified. As the object came closer, he locked it on radar and attempted to fire a heat-seeking missile, but all electronics went dysfunctional. Instrument dials turned round and round in confusion, and communication became difficult. Then the smaller object flew underneath the larger and rejoined it.

The shah of Iran took special interest in this report and visited with the pilots involved. He agreed with the pilots' assessment that the UFO was not of this world, and he informed them that this was not the first report of its kind.

This report is remarkable because an air-to-air missile launch was attempted by a human with hostile intent toward a UFO. It would appear that the UFO had some sort of jamming device and could sense when it was necessary to use it. But at the same time, the UFO showed no hostile intent and inflicted no harm.

THE PERUVIAN UFO INCIDENT

At 7:15 a.m. on April 11, 1980, a balloon-like object was observed at the end of the runway at the Joya Air Force Base in the southwest city of Arequipa, Peru, just southeast of the famous Nazca area. Commander Oscar Santa Maria Huertas scrambled his new supersonic Sukhoi 22 and attempted to approach and shoot down the object, as it was in restricted air space. He climbed to eight thousand feet and positioned for an attack. He fired a burst of

30 mm shells at the "balloon." The attack did no apparent damage. The object then flew away from the base at such a rapid speed that Commander Huertas had to fire up his afterburners and adjust wing parameters for supersonic flight. Doing so, he realized that this was no balloon. At a speed of 600 mph, the object maintained a position of about 1,600 feet in front of him. At an altitude of 36,000 feet, Huertas positioned for another attack, but just as he was about to fire, the object flew vertically and stopped above him. This happened three times. He increased his speed to 1,150 miles per hour, and the object still evaded him. He then found himself at an altitude of 63,000 feet, the upper limits of his plane's flight envelope. He attempted to position for another attack but could not; the object was just too maneuverable. Then the Sukhoi 22's low fuel indicator light went on, and he had to return to base.

Nonetheless, he had been able to get to about three hundred feet from the object and discovered that it was obviously not a balloon but a shiny object, about thirty-five feet in diameter, with a smaller, cream-colored dome on top. The base was silver colored and apparently metallic, with no visible means of propulsion. The pilot was terrified and thought that this vehicle, being of far greater technology, should easily take him out. After all, he was the aggressor. But the UFO did not attack, and Commander Huertas safely landed his plane.

THE JAPANESE AIRLINE FLIGHT 1628 UFO INCIDENT

During the night of November 17, 1986, a Boeing 747, JAL flight 1628, was flying from Paris, France, to Japan with a cargo of wine. There was a crew of three: Captain Kenju Terauchi, with twenty-nine years of experience; First Officer Takanori Tamefuji; and Flight Engineer Yoshio Tsukuda. Near the end of the Iceland-to-Alaska leg of the journey, Captain Terauchi noted a "very big" object to his left side that seemed to be following him. He later described it as a large walnut-shaped craft about two to four times the size of an aircraft carrier. With permission from the FAA air traffic controller in Anchorage, he changed altitude and course, but the large object followed him. He received permission to change flight altitude again, from thirty-five thousand feet to thirty-one thousand feet. Despite this change in altitude, Captain Terauchi reported that the craft still followed "in formation." The Military Regional Operations Control Center (ROCC) at Elmendorf then notified the air traffic controller in Anchorage they had a "flight of two" on their radar in the area of JAL 1628, confirming the captain's report.

The captain was given permission to do a 360 degree turn to the right to see if he could "shake off" the object. However, when he returned to his southward course, the object was still there on the left side, about eight miles away according to onboard radar. Captain

Terauchi made it very clear that he feared for his life. The object was tracked on the onboard radar for about thirty minutes.

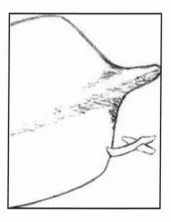

Artistic Representation
to show perspective

Two smaller crafts at one point emerged from the larger craft. They moved about very quickly and jerkily, stopping suddenly; they were very difficult to follow and obviously under intelligent control. The captain reported that the smaller crafts would go up to the left captain's window and shine a light into the cockpit so bright that the crew felt warm from the energy.

At 6:53 p.m., JAL 1628 notified air traffic control that they no longer had the object in sight. Thus, for about fifty-five minutes, JAL 1628 was accompanied by a large craft and two smaller ones obviously not from this

Earth and most certainly not made by Boeing. Clearly, this would classify as a near miss.

Captain Terauchi went public with this story and was promptly demoted to a desk (flight officers are not supposed to see UFOs). Fortunately, he was reinstated a few years later. He made his report to NARCAP (National Aviation Reporting Center on Anonymous Phenomenon) to Richard Haines, a retired NASA scientist. His report is now part of the public record.

By January 1987, word of this incident was filtering to the press, so Anchorage air traffic control notified John Callahan, the FAA Chief of Accidents and Investigations. Callahan told them to tell the press the matter was under investigation and to send all the data, including radar and voice recordings, military and civilian, to the FAA technical center in Atlantic City. Callahan's team then performed an analysis of the data and confirmed that a large object was indeed in the vicinity of JAL 1628, as described by Captain Terauchi. He then took the data to a White House briefing with three members of President Reagan's science advisory team, three members of the FAA, and three officials from the CIA. Callahan played the tapes about three times.

John Callahan was thinking that this object was some sort of new bomber or stealth craft from the air force, despite the fact that military radar in the vicinity of Anchorage denied any flights of any kind. Then one of the CIA representatives stood up and said, "This never happened. You were never here, and you are all sworn to secrecy," and confiscated all the data. John then asked the official, "Well, what do you think this is?" The CIA representative replied, "It's a UFO."

John Callahan later said that "they had all those people swear that this never happened, but they never had me swear it never happened." He also stated that "for those people who say that if these UFOs existed, then they would someday be on radar and that there would be professionals that would see it...back in 1986, there were enough professional people that saw it. What I can tell you is what I've seen with my own eyes. I've got videotapes. I got the voice tape. And I got the reports that were filed that will confirm what I've been telling you."

Voice tapes. Video tapes. Military and civilian aviation confirmation. Eyewitnesses with nothing to gain and everything to lose by going public. This is obviously compelling evidence that we are not alone in this world.

I know that President Ronald Reagan knew of this event and that it had a profound effect on his life and perspective of thought. On September 21, 1987, ten months after the JAL 1628 encounter, he gave an insightful speech to the Forty-Second General Assembly of the United Nations, a speech known famously as the "extraterrestrial speech":

In our obsession with antagonisms of the moment, we often forget how much unites all the members of humanity. Perhaps we need some outside universal threat to make us realize this common bond. I occasionally think how quickly our differences would vanish if we were facing an alien threat from outside this world. And yet, I ask you, is not an alien force already among us?

OTHER TESTIMONIES

Edgar Mitchell is no slouch. He has a PhD in aeronautics. He is a graduate from the Massachusetts Institute of Technology (MIT). He was the sixth man to walk on the moon during the *Apollo 14* mission. He has stated many times that "there have been ET visitations, and UFOs have been crashed" and studied. He insists that there is an organized policy of disinformation by the US government and that high-level elected officials are completely in the dark. If interested, you can read for yourself his book, *The Way of the Explorer.*

Vice Admiral R. H. Hillenkoetter was also no slouch. He was director of the CIA in 1960. He admitted to the existence of UFOs and also stated the air force "has silenced its personnel."

Colonel Gordon Cooper is another nonslouch. He was a *Mercury* and *Gemini* astronaut and witnessed a UFO that landed in full view, with landing tripod, at Edwards Air Force Base. He even took photos of the inside through a porthole. Then the craft flew away at "a very high rate of speed." You can read about it in his book, *Leap of Faith: An Astronaut's Journey into the Unknown*.

The list is almost endless. However, if you are still not completely satisfied, then I would like to direct you to the book *Disclosure*, by Steven M. Greer, MD, or to Greer's website, www.disclosureproject.org.

Conclusion

These are a few examples, out of many, that I thought were compelling. I believe that we can extrapolate at least four postulates from this data.

(1) UFO encounters are real and extraterrestrial.

(2) Extraterrestrials are very interested in the biological integrity of this planet and keeping Earth free from nuclear contamination.

(3) I suspect also we can deduce that they like French wine.

(4) The verification of the UFO phenomenon lends credence to the ancient Sumerian texts regarding ancient alien visitations and supports the belief that they never left.

CHAPTER SEVEN

THE NATURE OF THE GODS
The Great Secret

This concerns the nature of the gods, the noblest of studies for the human mind to grasp.

–Cicero, in *De Natura Deorum* (*Nature of the gods*), circa 50 BC

Weaving the Fabric of Reason and Digesting What We Have Learned

We have learned that, when one considers the cerebral perspective, there is a serious problem with the common belief in unorchestrated human evolution. The extremely rapid development of the human brain in such a short period of time raises the idea that something else happened, something profound, about four hundred thousand years ago. As if out of nowhere, suddenly we have a bipedal humanoid with a brain capacity very close to that of *Homo sapiens*. This bipedal is *heidelbergenesis*, which supposedly developed into *Homo sapiens* 250,000 years ago.

Often, we hear that our nearest living primate, the chimpanzee, has about 98 percent of our human DNA, and somehow that is supposed to convince us that the evolution of humans is just a short step beyond the chimp. Having 98 percent of DNA should not be too surprising. After all, doesn't our chimp brother have a spleen, liver, kidneys, heart, bone marrow, and intestinal track similar to ours? Doesn't he have ten fingers and ten toes? Binocular vision and bipedal ambulation similar to humans?

But our brain is much bigger—much, much bigger. In fact, our closest relative bipedal hominoid, *Australopithecus* (only 2 million years ago), had a brain capacity still only one-third of ours. Therefore, from a cerebral sense, this "short step" is a grand canyon, and thus we have missing links. Bright people are increasingly coming to the conclusion that animal DNA was manipulated by ancient extraterrestrials with a humanoid goal in mind.

We have also discovered that the biochemistry of human cerebral evolutionary embryological development was something we are only beginning to learn and what we are learning is stunning. Clearly and unequivocally, nature favored physical development over cerebral for survival. Genes responsible for cerebral embryological development such as HAR-1 (discussed in chapter 1) are very conservative and change only very, very slowly. Nature is not interested in our appreciating a Rembrandt painting or music from Mozart. Its top priority is survival, and thus it leaves aesthetics for much, much later. So, with no biological and almost no paleontological evidence of any preexisting similar primates, mankind, with his massive brain, suddenly presents himself some 250,000 years ago, to the Earth, and the Earth was never the same. We are not just different cerebrally but physically as well. We just do not naturally fit. Toilet paper is a constant daily reminder that the human is very different from our nearest relative and that we are not completely biologically amalgamated. All other animals live on an encounter basis. Humans alone exploit. Yes, I've seen chimps use a straw to get more ants from an anthill, but frankly (let's be honest), this is not very impressive. We humans must exploit our planet in order to fit in. We have to cut down trees to make toilet paper. We farm cotton and weave it into beautiful patterns for clothing. We need gloves and shoes for protection and warmth. We extract tin (a fairly complicated process) and mix it with copper to make bronze. Granted, we evolved from these lower animals, but this is clearly not the entire story. Hundreds of millions of years of cerebral evolution was drastically cut down to size by the introduction of ancient alien DNA into our bipedal relatives, and, like magic, humans started walking on Earth.

However, as discussed previously, the human literary record is not empty on this topic. When our ancient Sumerian relatives spoke of this, they described a story of hybridization, earthly animals manipulated with alien DNA and an *alien brain* for the purpose of serving the gods. We have examined the ruins of ancient structures dotting the landscape all over the Earth, which clearly resulted from highly sophisticated technology. We have also learned that the UFO phenomenon is real—compelling evidence that the Anunnaki are still among us.

We have learned that the origin of our consciousness is not our brain but a marvelous spirit that uses the human brain as a temporary integrator to make sense of our four-dimensional world. This compelling evidence for a soul also supports a belief in reincarnation.

And if we can discover evidence for spiritual forms of existence, then our creators most likely have working knowledge of this as well. This understanding provides us with a key to their motive.

However, before we weave all this together, let's examine the stories from ancient Sumer and how it relates to the Bible.

WHY ANCIENT SUMER?

Why ancient Sumer? Well, the Acadian language from ancient Babylon is the descendent language from the great Sumerian civilization. So we learned much of Sumeria from Babylon (which came later). In ancient

Babylon, provided you were educated, you would be able to speak the Sumerian language, much like the educated Europeans in the Middle Ages were able to speak Latin.

Everything that we take for granted in our modern world began in Sumer about six thousand years ago, on the shores of the Persian Gulf in what today is Iraq. The first wheel. The first laws. The first art and music. The first organized civilization with writing, commerce, and agriculture. The first multistory dwelling. The first running water and sewage system. Culinary skills, architecture, engineering, science and mathematics, woven fabric—it all started in Sumer. We get "ninety degrees to a right angle" from Sumer. We get "sixty minutes to an hour" and "sixty seconds to a minute" from Sumer. We get the twenty-four-hour clock from Sumer. As if from nowhere, suddenly Sumer pops up with all this sophisticated learning. Did mankind suddenly walk out of the Neolithic period and say, "Enough of this bullshit. Let's create a real culture"?

A brief Review of the Sumerian Account

So for a brief review, ancient Sumerian texts tell us that this planet was visited by beings from outer space hundreds of thousands of years ago, beings that wanted to exploit minerals from our planet. Our forefathers tell us that they were particularly interested in gold. Ask any aerospace engineer about the importance of gold, and he or she will tell you: it is critical. I suspect, however, that it was more than gold, as copper, tin, silver were certainly needed as well. The Anunnaki discovered that mining is backbreaking work (as it is today), and so Enki suggested that the mines would be worked by a creature created from the bipedal hominoids currently on Earth. They would be unable to multiply, so as to avoid adversely affecting the natural evolution of the planet. When they died, their genes would die also. They could never be accepted as equal with the Anunnaki, as DNA used in their construction came from the ancient rebel Kingu.

Enki's wife, Ninmah, was very educated in this type of enterprise and started to create the first intelligent worker beings, intended for mines in southeastern Africa. However, like all scientists, Enki and his wife went further and developed an advanced special human, the *Adama*, and moved him and his female companion from Africa to their beautiful home in Edin. The similarities with ancient biblical text are obvious.

We have also learned that our ancient Sumerian ancestry also had a Great Flood story that parallels the Bible account.

Is this some wild, concocted story, dreamed up by an ancient ancestor's creative imagination, or is there any truth to this wild tale? You might be surprised to find out that the Bible tells a similar story. So let us take a moment to see what ancient Hebrew scripture has to say on the topic of the nature of the gods. We will discover that the Bible, when carefully examined, also supports this ancient alien theory.

THE BIBLE
Genesis and Ancient Sumerian Text

And God said, "Let <u>us</u> make man in our image and after <u>our</u> likeness."
–Genesis 1:28

The Lord God took the man and put him in the Garden of Eden to till it and keep it.
–Genesis 2:15

The Hebrew word for *god* used in this text is *Elohim*, which in plural translates to *us*. The gods—plural—made us in their image and after their likeness. We look like them, but we were not supposed to think like them. The *Adama* was special and, thus, privileged to work in the gods' households and on the gardens at Eden.

However, something went terribly wrong. The Adama began to do things and say things that they were not supposed to do with a supposedly less-developed cerebrum. Worse yet, they began to procreate. They knew that they were naked and demanded clothes. They wanted utensils to eat with.

One thing we know about these Anunnaki beings, Enlil, Enki, and Anu, is that for them, ethics was the highest priority. These are not mindless beings bent on their own desires. These are highly intelligent beings with a standard code of conduct, either inherent or passed down from a higher being such as Anu. Thus, this story teaches us that their nature is one of justice and compassion.

Let us take a moment and examine the Anunnaki hierarchy.

THE NATURE OF THE TRINITY

Why the understanding of consciousness is so important to our quest.

When one studies the ancient Sumerian texts, it becomes clear that Enlil's personality is akin to the God of the Old Testament of the Bible. He is a stern disciplinarian and is very clear on matters of right and wrong. Enki (who loved to sail and fish) is more lenient with the humans, like Jesus in the New Testament; after all, they were his and his wife's creations. Anu is the Supreme Being and has final say. He is an all-powerful being that does not reside in the Earth's plane of existence but resides somewhere metaphysical and spiritual. He is akin to what Christians describe as the Holy Spirit (or what American Indians call the Great Spirit). Thus, the Sumerians teach us of a godhead trinity: the father (Enlil), the son (Enki), and the spiritual (Anu).

Zecharia Sitchin, author of *The Earth Chronicles*, reports that these beings, having mastered medical as well as physical science, live perhaps hundreds of thousands of years. This longevity is clearly part of the Sumerian record, as recorded in the "King List." But what about Anu's home planet, Nibiru, and its highly elliptical orbit? Isn't that a bit of a stretch? Although a small comet can maintain a highly elliptical orbit and stay in the sun's influence, astrophysics 101, on the other hand, clearly states that after perhaps only one "fling," something the size of a planet would gather so much speed and momentum it would be sent out of the solar system, never to return.

Thus, Nibiru is a place (not a planet) that does not follow the commands of our universal rules. It is part of the spiritual realm, not the physical. I believe anyone with an astrophysical background would pick that up. Zecharia Sitchin was a bright man, and I recommend all his books. But in terms of the "planet" Nibiru, I believe he was wrong.

Thus, the first two gods, Enlil and Enki, are real physical gods, but Anu is the big guy in the spiritual realm. This would make sense, because even if you can live for hundreds of thousands of years, you will die eventually. And at that point, your soul (the real you) will enter the realm of Anu, and your eventual disposition (reincarnation, etc.) will be up to him and his consort. Thus, Anu holds the key and has the last say. One does not mess with Anu. He's the big one, and if we can begin to understand this spiritual knowledge, in our relatively primitive condition, the Anunnaki most assuredly have an intimate familiarity and perhaps even a working knowledge of it as well. Understanding the origin of human consciousness is critical, if we are going to begin to understand their world.

SOMETHING WENT TERRIBLY WRONG

Anyway, the new humans were *not* supposed to procreate and become part of Earth's evolutionary program. They were *not* supposed to think on their own but rather communicate and receive instructions for work detail. However, either by further gene manipulation or by accident, the new humans began to think like the gods.

Now we have the ultimate ethical dilemma. What is to be done with the humans, now that they have become intelligent, procreating souls? This raises the interest of Anu, for his business is souls.

Let us go to the Mosaic text to help unravel this story. Considering this multigod physical/spiritual concept, there are some things in ancient biblical scripture that become very enlightening.

> *And they heard the sound of the Lord God walking in the Garden in **the cooler portion of the day.** Now Adam and his wife hid themselves among the trees, from the presence of the Lord. The Lord called out, **"Where are you?"** And Adam said, "**I was afraid***

because I was naked," *for they have sewed fig
leaves together to make a loincloth.*
      –Genesis 3:8

There are several hints tucked away in this verse
of scripture. The *Lord God walking in the garden in the
cooler portion of the day* denotes a physical god. One
subject to things like heat and humidity. A physical god
that can become uncomfortable and display a preference
for cool temperatures. A god in the spirit form would not
display such a preference.

The Lord God then called out, "Where are you?"
demonstrating a physical limitation of sight and perhaps
hearing. A spiritual god would not have such a limitation. One can hide only from a physical god.

The statement from Adam that "I was afraid because I was naked" not only denotes that he is now cerebrally more advanced than desired but also suggests that
the Lord God was physically clothed. (Why would Adam
seek to cover his nakedness, unless that was behavior modeled by his god?) A spiritual god does not need
clothes, but a physical one would.

Thus, the Lord God as described in the third chapter of Genesis is a real, physical god of flesh and blood. A
god that can get too hot or too cold. A god one can hide
from. A god wearing clothes.

So Adam and Eve were cast out of the garden and
deprived of the Tree of Life. They were to be on their

own. This raises the question: Why? What did the gods have to fear?

The key to this puzzle is the DNA. The "divine" DNA of the "gods" could not be polluted by the earthlings. What kind of offspring would be produced? Would they be rebellious, cruel, and immoral? Insight comes from the Enuma Elish, where it clearly states that blood (DNA) from that great rebel Kingu was used to create mankind. Might they reproduce exponentially and threaten the ruling class? Mankind had to be cast away, and the gods (the Anunnaki) were instructed *not* to have sex with them. This is reminiscent of the New Testament text.

> *All have sinned and come short of the glory of God.*
>
> –Romans 3:23

An interesting footnote occurs in Genesis 3:20, when God makes clothes out of animal skins for Adam and Eve. He could have given them the finest clothes in his closet, but rather it was animal skins. He wanted to show them how to live and survive in the wild. This is an indication that the gods have visited mankind many times in the past and looked after us, teaching us the use of herbs and the skills of husbandry and farming. The physical evidence is there as well, considering the ubiquitous Stonehenge-like structures dotted across the earth from Armenia to

Great Britain. To the Wall Street bondsman, knowing the winter solstice is immaterial, but to an agricultural society depending on crop success, it is life and death. Note, then, the Temple of the Moon in Cusco, Peru, which was designed with the winter solstice in mind.

The humans had some help.

Nonetheless, we shall see that it was *not* the Anunnaki DNA that was polluted, but rather that of the new humans.

SEX AND VIOLENCE

When people began to multiply upon the earth and daughters were born to them, the sons of God saw that they were fair and took wives for themselves of all that they choose.

–Genesis 6:1–2

Interestingly, the Bible also states that the *Nephilim* (Hebrew for "those who from the heavens came") were "upon the earth in those days *and afterwards too*" (Genesis 6:4). In other words, they were here in those days and—*oh, by the way—they never left.* I suspect that this is a subtle hint that the UFO phenomenon is the Anunnaki...

While the Bible gives us the *Reader's Digest* version, the Book of Enoch tells us a much more complete story. Remember, Enoch was the seventh patriarch (king) before the Great Flood. The Bible said that he "walked with the *gods* (Elohim)" (Genesis 5:24), and he was so good that after 365 years of life, God took him. I guess his DNA was good enough for the gods. Enoch even gives us the rebels' names and the name of their leader (Samyaza). He tells us that two hundred of the rebels descended upon Mount Herman, located today in southern Lebanon near the border of northern Israel. Enoch even gives the names of the lieutenants—Urakabarameel, Akibeel, Tamiel, Ramuel, Danel, Azkeel, Saraknyal, Asael, Armers, Batraal, Anane, Savebe, Samsaveel, Ertael, Turel, Yomyael, and Arazyal (Genesis 7:2).

The book of Enoch then goes on to describe the horrible nature of their offspring. These were greedy, cruel people, and they began to devour the Earth's resources. This suggests that making a moral, compassionate man (Adama) is much more intensive and difficult than making just an intelligent being.

> *Now the Earth was corrupt in God's sight, and the earth was filled with violence.*
> –Genesis 6:11

> *The Lord God saw that the wickedness of mankind was great and that their every thought was continuously evil. And he was sorry that he made mankind... and said, "I will blot out the human being from the earth that I have created."*
> –Genesis 6:5-7

The Babylonian text *Atrahasis* describes two gods of interest in the flood story, Enlil, the disciplinarian, and Enki, the compassionate. Sumerian texts report that Enlil was tired of the "noise"—no doubt constant reports of their evil, murderous deeds—and so he intended to wipe them all out in a great flood. He then forced Enki to take a sacred oath not to the expose the plan to the humans. Enki became distraught because he did not want to destroy the good along with the bad. After all, some of the humans had pure, unpolluted DNA and were worth saving. For this reason, Enki passed information to Atrahasis (the biblical Noah) through a reed wall. (This avoided breaking his oath because the "conversation" was overheard and not directly transmitted.) There is an ancient clay tablet picture of Enki talking though a reed wall, with Atrahasis listening on the other side. Atrahasis follows the advice from Enki, builds a great boat, and saves his family, as well as many animals. The key as to why Atrahasis was spared in the Great Flood is given in the Bible.

It should be noted here that there is at least some evidence for a great flood in the past. The last ice age ended fairly quickly about fourteen thousand years ago, and atmospheric temperatures were rising rapidly, resulting in violent climatic changes. This rise in temperature allowed for a gradual melting of the Antarctic ice caps. Remember that the Antarctic ice caps sit on land, and when the Earth's temperature increases, a muddy, slippery floor is created, separating the colossal ice cap above from the earth below. This results in slippage of the ice cap and its collapse into the ocean, creating a colossal tidal wave of biblical proportions as well as increasing the surface level of the ocean.

However, it is not just the Bible that describes such things. Such an event is also described in *Scientific American* (March 1993) and the prestigious journal *Science* (January 1993). Thus, a giant tidal wave, as high as 1,500 feet, is estimated to have come from the south, through the Persian Gulf and ancient Sumer, and to the mountains of Ararat in what is today Turkey.

AND NOAH WAS PURE IN HIS GENERATIONS
–Genesis 6:9

The Bible also goes on to explain that Noah was a direct descendent from Adam and also was the tenth patriarch. In other words: he had good DNA. There were no violent demigods in the Adam-to-Noah line of generation, and thus it was something worth saving.

After the flood, Enlil was angry that Atrahasis was saved. But later he was grateful and thanked Enki for saving his genetic lineage, which later populated the Earth. Or was the genetic component tweaked or altered—by Enki, Anu, or someone else—just prior to the flood?

For this, we go back to the Book of Enoch (chapter CV), which tells the story of the birth of Noah. Noah's father, Lamech, was surprised to find that his new son looked very much like the watchers (Anunnaki). His hair was white and full, his eyes were beautiful, and he talked at an early age. But Lamech's father, Methuselah, assured him (through Enoch, who was living among the gods by that time) that the child was indeed his. At least she wasn't fooling around with the pool boy.

Thus, it would appear, a special gene for the population of the world was formed. A gene not polluted by the mixture of humans and Anunnaki. A gene good enough to go it alone. Good enough for the development of souls—although a *guide* might be needed every thousand years or so. Good enough to understand DNA and its influence on human behavior (provided that the DNA was a stable factor and not polluted by genetic aberrancies).

DID THEY KILL THEM ALL?

But was the flood complete in the destruction of all the genetically aberrant demigods? Did the Great Flood get them all? The answer is given again in the Bible (Numbers 13:33). Moses had brought the children of Israel to the Promised Land and sent out spies to survey the enemy. However, the spies reported that the men were too large and too strong, and so the Israelites were fearful and went back into exile in the desert. The key is the fact that they reported the presence of the Nephilim (those who from the heavens came).

And so the flood did *not* destroy all the demigods. Furthermore, the Numbers biblical text goes on to say, basically, "Oh, by the way, the Anunnaki that you heard about from the ancient Sumerian text? Well, they are the same as the Nephilim that we talk about in the Genesis record." It's all there for you to read. This explains why God ordered some cities to be completely destroyed—every man, woman, and child. It was simply divine eugenics. The experiment was not to be screwed up. (God can do that. He is, after all...*God*). This slaughter was simply to help complete the work of the Great Flood.

This, of course, begs the question: *For what purpose, and to what end*?

Let Us Examine the Evidence So This Is Where We Stand

Incredible as they may seem, certain realities are inescapable. They won't simply go away, and these facts must be entered into the formula of existence and our understanding of the nature of the gods. Any theory of existence without them would be meaningless and irrelevant.

I am suggesting that human evolutionary development was artificially accelerated for purposes outlined in ancient scripture.

1. Primate DNA was manipulated by extraterrestrials to develop a primitive worker, and this resulted in the development of humans.

2. Higher intelligence was involved in the construction of ancient archeological sites.

3. The origin of human consciousness is spiritual and not cerebral. You are a spiritual person in a temporary human form.

4. Reincarnation is a fact and is most likely part of a plan for self/soul development.

5. Ancient inscriptions give clues to our past and are not in conflict with ancient Hebrew scriptural text.

6. The UFO phenomenon is real and consistent with the above evidence of alien presence and influence and supports the belief that the gods never left.

7. The aliens have a thorough understanding of the physical and spiritual universe. In their world, religion is not required. Fact replaces faith.

The evidence proving these conclusions is compelling and cannot be ignored or swept under the carpet (as commonly done by mainstream scientists, for fear of professional retribution or ridicule). They must be taken as what they are: relevant facts that should be elemental parts of our understanding of existence and the nature of the gods.

LET US EXTRAPOLATE FROM THE EVIDENCE
"Come let us reason together." (Isaiah 1:18)

The god of the Mosaic text is a real, physical god. A god of flesh and blood. Not a nebulous, faraway person with little interest in the affairs of the world and the suffering of mankind. He is interpreted as a "supreme" god due to the obvious and extreme technological gap, but nonetheless, he is an ancient alien.

OTHER ANCIENT TEXTS REGARDING THE NATURE OF THE TRINITY

There are three primary figures in the godhead that form a *trinity*. Enlil, the disciplinarian, is the main Earth god, the God of the Old Testament. The Book of Enoch describes him as the "Ancient of Days." The Zohar (sacred writing of the Jewish kabbalah sect) states that it was the Ancient of Days who provided manna for the children of Israel during the desert wandering.

One might say that Enki, the compassionate, is the god of the New Testament. The Book of Enoch describes him as the "Son of Man," which, interestingly, is the same description given to Jesus in the New Testament. This is significant to our understanding, for it strongly suggests that Jesus was familiar with the Book of Enoch, as was Daniel (see Daniel 7:13-14).

Then Anu, the sky god, is the Holy Spirit, the all-knowing and all-powerful god who has the last word in any major decision. Little children and thousands of adults who have experienced near death have stated that they were in *heaven* and not any other place. They were in a spiritual place, suggesting that the supreme god is a spiritual god.

Thus, we have sacred and secret information given generation after generation: a trinity of a godhead. Enlil, Enki, and Anu: Ancient of Days, Son of Man, and the Lord of Souls. All sewed together in a perfect fabric. With such a trinity, all aspects of existence are covered—the physical (Enlil), the spiritual (Anu), and the communication between the two (Enki). The plan is perfect and perfectly consistent with Jewish and Muslim monotheistic traditions.

The origin of human thought is not the human brain. The human brain is an integrator. Consciousness originates from the spirit residing in the universe, of which space and time are more like illusions, not fundamental principles of existence.

THE TRUE NATURE OF BEING

Because the origin of human consciousness is spiritual, it can pass on from life to life. It can reincarnate from human to human, to learn and become wiser and more understanding, as well as compassionate. This forms the learning curve of existence and soul development.

This is dependent on a proper physical substrate. Here on Earth, the physical substrate is the human animal, which must have a proper balance of *yin* and *yang,* good and bad. With this balance, the soul is free to decide its own path and is not overly influenced by primal physical urges that originate from aberrant DNA. Thus, human life has been orchestrated to serve a higher purpose.

We somehow came to believe that God is in love with humans. Perhaps he is. But humans may be just tools, a means to an end, bacteria in the petri dish. Are we humans simply a means to find a more perfect and balanced physical cerebrum to amalgamate the true origin of consciousness with the physical form? Anu may or may not be in love with the human race. Let's not forget that ancient Sumerian text tell us that the genes that were amalgamated into our DNA came from the rebellious Kingu, no great friend of Anu.

One may also conclude that because humans are not part of the "family," we require an intercessor from time to time, an advocate to explain our behavior. And so I suspect Enki, who clearly was in love with the human race, was reincarnated as a human many times (as Jesus, Mohammed, Confucius, Socrates, etc.), not only to show us the way but also to experience our biological drives and understand, through that experience, our sometimes awful behavior.

WHAT ABOUT THE UFO?

The UFO phenomenon is real, and although there are many hoaxes, the basic premise that there are extraterrestrials in constant contact with us is not fallacy. Moses in Genesis said they never left. The Book of Enoch talks of them. The ancient Sumerian scriptures talk of them. In fact, as we have learned in the sixth chapter, there is a mountain of credible testimony—including radar confirmation—pointing to their existence... but why don't they reveal themselves?

I believe the most logical conclusion is that they do not contact us directly, because they are under orders not to. They, the ruling body, want to know how this set of genes pans out, and the only way is by letting us live on our own. In the ancient alien theory community, this is referred to as the "zoo hypothesis," and I believe this makes the most sense when considering all the evidence. The occasional sightings are probably errors or mishaps involving the cloaking and stealth usually maintained by the watchers.

As indicated in the Sumerian text, the owners and perhaps operators of these crafts, the Anunnaki, have amazing life-spans, thanks to their extensive knowledge of medical and physical science. Their crafts obviously operate by the use of *unified field theory*, the greatest enigma of modern physics. It is a theory that unites the world of the electromagnetic spectrum (light and radio waves) with that of gravity and time, through mathematical formulation. In other words, the Anunnaki can manufacture gravity for flight, as well as for personal comfort and safety inside the craft. This allows the crafts to go extremely fast and take incredibly sharp corners without injuring or even disturbing the pilots.

In his books *Pyramid Wars* and *Divine Encounters*, Zechariah Sitchin talks of rockets and rocket bases. This is where I have to differ. I can guarantee you that alien visitors, ancient or otherwise, did not come to Earth on crude rockets based on primitive Newtonian physics. Clearly, the Anunnaki are much more advanced, beyond our understanding of the physical universe. This is a culture so scientifically advanced that we simply would have tremendous difficulty relating to it.

What is clear from the ancient Sumerian texts is that there's a sort of divine trinity that is very reminiscent of the same trinity given in the Book of Enoch, the Book of Thomas, and even the New Testament portion of the Bible. I suspect that the order for *watching only* comes from this ruling body.

LET US EXTRAPOLATE FURTHER

I suspect that souls in human form are here by choice. (Dannion Brinkley was told this during his incredible near-death experience.) They are exposed to a physical and sometime cruel existence to learn and become wiser, more compassionate, more tolerant, and more understanding of other beings. By doing so, they become closer to God (Anu). In addition, their shorter Earth-based life-spans allow quicker development of the soul through reincarnation.

The spiritual soul must be exposed to a physical world but in proper balance. The ancient Egyptians called this the Ka and Bah souls. The Ka was the physical, flesh-and-blood soul—or the hardwiring of the brain, as we tend to think today. The Bah soul is the spiritual soul that reincarnates. The physical hardwiring of the brain comes from our DNA, and the early DNA that formed the antediluvian demigods was much too violent. For them, spiritual growth was impossible, and so these genes must be removed from the human gene pool in order to form a proper balance. Just imagine being reincarnated into a Tyrannosaurus rex. You're not going to go out into the field and pick daisies. You're going to do what a Tyrannosaurus rex normally does, and that is *kill*. Not much room for spiritual growth.

ARE THERE BAD ANUNNAKI?

Not all the Anunnaki are good. The Bible and ancient Sumerian texts make it very clear that God has to go to war from time to time. He is not going to war against the good guys. However, I think you can bank on the fact that he is going to win (I certainly hope so). This Earth, by forming good souls, helps reinforce his army. This is why the bad Anunnaki are hateful, jealous, and resentful of human souls that are demonstrating good behavior.

A cuneiform fragment discovered in southern Iraq in the summer of 2006 describes "the mighty men of renown who expelled the Nephilim." Obviously, if it is *good* that someone is expelled, then that someone must have been bad. The Roman Emperor Marcus Aralias once said, "There is always someone else to fight."

DO THE ANUNNAKI HAVE ANY RELIGION?

Given the obvious fact of the Anunnaki's technical, scientific, and medical advancement, it would be logical that they would have a much higher understanding of the nature and origin of consciousness. They would also have a much higher understanding of what we would call "spiritual matters" and a better relationship with the supreme spiritual being, Anu. One should therefore conclude that they are *not* religious (in a human sense), as religion requires faith, and faith is no longer needed when one has full knowledge of and familiarity with spiritual matters. One does not need faith to know the current president of the United States; it is simply fact. The same is true of the Anunnaki when it comes to Anu and the spiritual world.

WHAT IS IT THAT THEY WANT?

Stephen Hawking once said that the day we are visited by extraterrestrials will be a bad day for humans. This makes perfect sense. What do we have to offer these supreme beings? We make lousy company, as we are so ignorant. They won't eat us, as we have little muscle mass (and beef and pork taste better). There is nothing we can teach them. We live on a beautiful planet (that, by the way, we are fucking up with overpopulation, pollution, possible nuclear catastrophe, and violent crime). We are mostly (with very rare exceptions) self-centered, greedy, immoral, covetous, dishonest, lustful, and even religiously fanatical, and we have developed a love for violence. We are, simply, dangerous. What could we possibly offer them?

The only things most can imagine are servitude and sex, and I suspect that is not far off the mark. But there's another thing they are very interested in.

This is a very beautiful planet, with beautiful beaches and mountain ranges, and any true-blue platinum-blooded alien would jump at the chance to take charge of it. It would be easy to insert a virus that is specific to humans and simply kill us off. Who wants to fly around, hiding in a spacecraft for ten thousand years? It would seem easy for them to say, "Fuck the humans. Why should this beautiful planet be wasted on them? Let's go skiing, sailing, and fishing. Let's have a barbecue on the beach. Let's drink some wine, have a steak dinner, and have sex. Let's bring the wife and kids. Maybe a few hot-looking babes."

They haven't done it yet, at least not in our lifetime, so they must be under orders not to interfere. If I were them, I would definitely have to have some orders not to interfere. The temptation to set up a palace and have the humans serve me (after all, ancient Sumerian texts and Hebrew scripture both state that humans were made to serve the gods) or simply destroy them and repopulate the Earth with my own kind would be just too overwhelming. Hell, I get claustrophobic just thinking about it.

They don't interfere, because humans are an experiment. Here again is the zoo hypothesis, or the "petri dish hypothesis." The aliens are watching to see how this DNA pans out through human behavior. Some people criticize the zoo hypothesis, because, they say, it simply won't work. All it would take is a few observers (aliens) to muck it up by having sex with humans and changing the DNA pool.

The Book of Enoch (as we discovered in the second chapter) says that is exactly what happened thousands of years ago. Two hundred "gods" came to Earth, had sex, and literally fucked it up, albeit temporarily.

COCREATING WITH GOD

During his most incredible near-death experience, Dannion Brinkley (author of *Saved by the Light*) was told that souls who *volunteered* to go Earth in human form were particularly brave. He was also told that we humans are cocreating with God. The next logical question is: cocreating what?

Everything here is expendable. We all die. The great monuments to our civilization rust and degenerate. Even the *Titanic*, over two miles underwater, is disintegrating. Everything on this planet dies. This is a planet of death. What could we possibly be co creating?

The one and only thing that lives on is this expansive human, genetic, germinal layer of cerebral embryobiologic development. Always growing. Always there. The Ancient of Days (and his clan) prefers to exist in the physical form, and in order to do so, he (and she) will need a suitable DNA substrate to reincarnate into. This is vital. We are here in human form, good or bad, to help develop that suitable DNA substrate, to help determine the best DNA for the physical experience. If you are a god whose goal is to reincarnate, then you have a particularly keen interest in DNA. What are you going to look like? Are you going to have any personality deficits? Any medical conditions? There is a lot of stuff to consider with this DNA business.

Our benefit, as ordinary humans, is developing the soul through experience. When we die (and we are all close to death), we transcend and become closer to the big guy, Anu.

DNA IS KING
Welcome to the Farm
The meek will inherit the earth.
–Matthew 5:5

In the world of the Anunnaki, DNA is king. The Anunnaki live in physical form, and although they have fantastic life-spans, they, too, realize that death is part of their future. However, with their super-advanced scientific and technical achievements, it should *not* be too much of a stretch to conclude that they would want to be able to influence their long-term future through reincarnation. In this way, good DNA is king, and to determine good DNA, one must make changes and then watch how it develops over many thousands of years.

We must understand that "good DNA" does not mean hair color or cheekbone position. The concern is psychological drive and psychiatric disease. The more we learn about DNA, the more we begin to understand the power that genetic influence has on behavior.

Let me give you one example. What greater drive is there than sex? Scientists can now create homosexual male rats by manipulating brain chemistry during their third trimester of cerebral development. Behavioral genetics is a new and rapidly advancing field of science. Now that the human genome is mapped out, and technology is rapidly developing machines to identify genes in mere minutes (instead of the weeks it used to take), the field is on the verge of "going viral." Genes for aberrant sex drives such as necrophilia and pedophilia and for psychological disorders such as schizophrenia, bipolar disorder, kleptomania, psychopathic, and emotional discontrol give us a deep appreciation for the power of behavioral genetics.

Thus, like it or not, we humans are part of a grand experiment necessary to develop that good DNA. Like it or not, we are part of the age-old debate of environment verses heredity. We are helping to discover the balance of DNA and environment that helps develop a better physical world.

I don't mean to insult the human race by implying that we are all being used by ancient aliens. The fact is that you and I are not heterosexual or homosexual by choice; it is simply the way our brains are hardwired. These neuronal circuits *can* dictate the behavior of the spirit housed in the brain. Furthermore, we can conclude that the *real* you is *not* human at all. You are a spirit, temporarily in a human form. So don't be too alarmed if you realize that the alien gods may want to do a bit of "pruning" from time to time to perfect human DNA.

At the moment, you are helping to collect data. Very valuable data that will help create a new world and a better human. You are doing your part, be it good or bad. Even the serial killer and serial rapist, as bad as they are, are helping develop an understanding of DNA and environmental influence on behavior.

If this is true, then is it not much of a stretch to think that the same process is going on in other parts of the universe, or perhaps even all over the universe? (Near-death patients of mine have told me this is true.) Aliens are collecting a mountain range of animal behavior data and their genetic/environmental influences to help determine a proper balance. A combination of good DNA and good spirit. A proper mixture of yin and yang.

And, as in all experiments, *there is an end*.

Is God in Total Control?

It must be noted that many great religions on both hemispheres believe that "God is in control" of all things. As a
physician, I have heard this many times enunciated by cler-
gy and loved ones comforting the sick and their families.
They believe that everything that happens on this planet is
the result of divine consent. Whether it is winning the lot-
tery or getting terminal cancer, it is all the "will of God."

The information presented in this book suggests that
this is not exactly so. It is not that simple. We are in control
of much of our fate. We are responsible for many of our mistakes. The smoker's lung cancer is not God's will. Although
we will determine the success or failure of our species, we
cannot ignore the powerful influence of behavioral genetics. Perhaps one might say it is the role of the genetic dice
that is "the will of God." But all of us, through reincarnation, get our chance to throw the dice, and in the end, with
enough varied experiences, it all balances out.

Regardless of the source of the drive, be it biological, environmental, or spiritual, data is being compiled
by our spiritual and physical "brothers." The Anunnaki
are observing and will not interfere as long as we don't
do serious harm to *their* planet.

As a physician working for decades in the trenches
of medicine, I have seen examples of great love and great
moral bankruptcy. On the whole, I believe, we are doing
fairly well, considering the stress of this environment.
At the same time, we must also consider our deplorable

record—thousands of years of constant violence to one another.

It is no secret that violence sells. If not in the Roman arena, where death became entertainment, then in movies and video games. There is a rebellious and violent biological drive in us that we must try to overcome. I suspect that powerful genetic-driven behavior originates with the DNA influence of that great ancient Anunnaki rebel Kingu, who tried to overthrow Enlil (and, I suspect, dared to even overthrow Anu) and was defeated by Marduk (a son of Enki). Remember, it was the blood of Kingu that Enki and Ninmah used to eventually make humans. I believe this ancient tale explains the source of some of our passion for violence and why we, as humans, may need a spiritual intercessor to explain some of our bad behavior to those in the spiritual world—what politicians today would call a *spinmeister*.

I do believe that currently there might be enough data to formulate a proper physical being worthy of a long and useful life on this planet. God may not be in total control, but he (or she) *is* the ultimate authority. We now know that genes can affect sexual drive, kleptomania, schizophrenia, Asperger syndrome, bipolar disorder, etc. Through your behavior, whether good or bad, you are doing your part as a temporary human being. You are helping to formulate the proper genetic code for a proper physical existence on this planet. Yes, you can call this eugenics, but don't forget: it is *divine eugenics*.

All of us are precious souls having a temporary human experience in order to cocreate with God. Just as there is an end to our physical lives, there is an end to this grand experiment.

However, is it just an experiment? I equate it more to a *farm*. Bad DNA is to be weeded out, and the good harvested. You might call it a DNA farm. (Perhaps there are many in the universe.) This is an ongoing process, growing and getting better. It is difficult for us to understand, with our short life-spans and limited knowledge of the real world around us—much like a pet dog, who has no concept of the Americas or Europe, no understanding of the real world around him, nor of its history, no understanding of purpose, living on an encounter basis, waiting for food, water, and the master to come home.

We humans have been shielded from the real world and its purpose. Someday this word will change into a much better place. A physical world triumphant. A physical world without rape and murder. A world without mindless suffering and cruelty. I believe that you and I are doing our part (perhaps a bit unwittingly) to make this happen.

Let Us Now Look at the Three Great Questions of Philosophy
Where Did We Come From?

We came from a combination of earthly evolutionary DNA and ancient alien DNA. Hence, we are hybrids.

Why Are We Here?

We are here to grow and develop souls for God and also to help develop the ideal DNA for physical existence. There will be latency until they (the Anunnaki) reveal themselves. The grand experiment is not quite complete. This DNA farm is not quite ready for harvest. They need to know if humans will eventually destroy themselves as they become more powerful with thermonuclear weapons and careless environment stewardship, not to mention religious extremism, selfish violent behavior, and social/racial intolerance.

Hence, we are cocreating with the gods.

Where Are We Going?

We are going back to God and to all our old friends... someday in the physical and most assuredly in the spiritual, but we are all going home...

Erich Von Danikin, author of *Chariots of the Gods*, was once asked if this revelation of ancient alien theory conflicts with his being a Catholic and his belief in God. His answer was, "No; I never lost my God nor my faith, and, in fact, this acknowledgment reinforces my belief and made God so much bigger." I agree with his assertion.

Science, our ancestors, and the stones are shouting at us; please listen.

**MAY THE GOOD LORD SHINE HIS LIGHT
ON YOU.
MAY EVERY SONG YOU SING BE YOUR FA-
VORITE TUNE.**
–The Rolling Stones

I hope this book has been inspiring and that you are a better person for it. And when it is time for you to pass on to the other side, may you be rocked like a baby in the powerful arms of God.

I would like to thank my special old friends for their assistance with the formulation of this book. The code is cracked. Enjoy.

David D. Weisher, MD

Bibliography

Laurence, Richard, trans. *The Book of Enoch the Prophet*. San Diego: Wizards Bookshelf, 1995.

Sitchin, Zecharia. *Divine Encounters*. Avon Books, 1996.

Butler, Alan, and Christopher Knight. *Civilization One*. London: Watkins Publishing, 2004.

Silverberg, Robert. *Lost Cities and Vanished Civilizations*. New York: Bantam Pathfinder, 1962.

Ehrman, Bart D. *Lost Scriptures*. Oxford University Press, 2003.

Scholem, Gershom. *Kabbalah*. New York: Dorset Press, 1987.

Kean, Leslie. *UFOs*. New York: Three Rivers Press, 2010.

Pagels, Elaine. *The Gnostic Gospels*. New York: Vintage Books, 1979.

Weisher, MD, David D. *Mysteries of Consciousness*. Xlibris Press, 2005.

Leininger, Bruce and Andrea. *Soul Survivor*. New York: Grand Central Publishing, 2010.

Brinkley, Dannion. *Saved By The Light*. New York: Villard Books, 1994.

Iverson, Jeffrey. *More Lives Than One*. New York: Warner Books, 1976.

Hancock, Graham. *Fingerprints of the Gods*. New York: Crown Publishers, 1995.

341st SMW Unit History, pp. 32-34, 38 (Freedom of Information Act)

www.topsecrettestimony.com/Witnesses/All Witnesses/Captain Robert Salas/tabid/290

Warren, Larry. *Left at East Gate*. 2005.

Vague OVNI sur la Belgique (UFO Wave over Belgium) SOBEPS organization

Hynek, Dr. J. Allen, Philip J. Imbrogno, and Bob Pratt. *Night Siege: The Hudson Valley UFO Sightings*. Ballantine Books, 1987.

Clarke, David. "The Rendlesham Files." http://drdavid-clarke.blogspot.com.

Bruni, Georgina. *You Can't Tell the People.* Sidwick & Jackson.

Federal Aviation Administration. "Chronological Summary of the Alleged Aircraft Sightings by Japan Airlines Flight 1628." January 6, 1987.

Maccabee, Bruce. "The Fantastic Flight of JAL 1628." *International UFO Reporter,* Vol. 12, No. 2.

Bahn, Paul. *Lost Cities.* New York: Welcome Rain, 1999.

Bauer. *Ancient Cuzco.* University of Texas Press, 2007.

"On Reconstructing Tiwanaku Architecture." *The Journal of the Society of Architectural Historians.* Vol. 59, no. 3 (358–371).

New Scenarios on the Evolution of the Solar System and Consequences on the History of Earth and Man. Milano and Bergamo, June 7–9, 1999. Universita degli Studi di Bergamo, Quaderni del Dipartmento di matematica, Statistica, Informatica ed Applicazion. Serie miscellanea, Anno 2002, N. 3, pp. 171–203.

King, L.W., trans. *Enuma Elish (The Epic of Creation)*. London, 1902.

Dalley, Stephanie, trans. *Atrahasis: Myths from Mesopotamia: Gilgamesh, The Floodland, and Others*. New York: Oxford University Press, 1991.

The Babylonian Gilgamesh Epic. New York: Oxford University Press.

Lambert, W.G., and A.R. Millard. *Atrahasis: The Babylonian Story of the Flood*. Eisenbrauns, 1999.

Dunn, Christopher. *The Giza Power Plant*.

Penrose, Roger. *The Emperor's New Mind*. Oxford, 1989.

Hameroff, Stuart. "Quantum Computing in Microtubules." *Japanese Bulletin of Cognitive Science*, 4 (3): 67–92.

Leick, Gwendolyn. *Mesopotamia: Invention of the City*. London and New York: Penguin Press, 2002.

Kramer, Samuel Noah. *The Sumerians: Their History, Culture, and Character*. University of Chicago Press, 1963.

"Scientists Identify Gene Differences between Humans and Chimps." *Scientific American*, August 17, 2006.

"An RNA Gene Expressed During Cortical Development Evolved Rapidly in Humans." *Nature* 443 (7108): p. 167–172.

Crick, F.H., and L.E. Orgel. "And Further Reference." *The FASEB Journal.* January 1993. 7 (1): 238–9.

Mullin, Glenn H. *The Fourteen Dalai Lamas: A Sacred Legacy of Reincarnation.* New Mexico: Clear Light Publishers, 2001.

Van Lommel, P. "Near-Death Experience in Survivors of Cardiac Arrest: A Prospective Study in the Netherlands." *Lancet,* 2001, 358: 2039–45.

Tucker, Jim. *Life Before Life: A Scientific Investigation of Children's Memories of Previous Lives.* 2005.

Stevenson, Ian. *Twenty Cases Suggestive of Reincarnation.* University of Virginia Press, 1980.

Schock, Robert M., ed. "Redating the Great Sphinx of Giza." *Circular Times,* 1992.

Gastaut, H., and J.S. Meyer, eds. "The Pathogenisis and Topography of Anoxia, Hypoxia, and Ischemia of the Brain in Man." *Cerebral Anoxia and the Electroencephalogram* (1961): 144–163.

Penfield, W. *Mystery of the Mind*. Princeton University Press, 1975.

Penfield, Wand, and Roberts. *Speech and Brain Mechanisms*. New Jersey: Princeton University Press, 1959.

Head, Joseph, and S.L. Cranston. *The Phoenix Fire Mystery*. New York: Julian Press/Crown Publishers, Inc., 1978.

Plum, F., and J. Posner. "Multifocal, Diffuse, and Metabolic Brain Diseases Causing Stupor and Coma." Chapter 4 in *The Diagnosis of Stupor and Coma*. Philadelphia: F.A. Company, 1985.

Weiss, Brian. *Many Lives, Many Masters*. New York: Fireside Books, 1988.

Fang, J.F., R.J. Chen, B.C. Lin, Y.B. Hsu, J.L. Kao, and M.F. Chen. "Prognosis in Presumptive Hypoxic Ischemic Coma in Non-Neurologic Trauma." *Trauma*, 47 no. 6 (1999): 1, 122–5.

"Central Nervous System Resistance: The Effects of Temporary Arrest of Cerebral Circulation of Periods of Two to Ten Minutes." *Neuropathology Experimental Neurology* 5 (1946): 131.

Von Daniken, Erich. *Chariots of the Gods*. New York: Berkley Books, 1999.

De la Vega, Garcilaso. *The Royal Commentaries of the Incas*. Urocyon Books.

Delgado, Jorge Louis, and Mary Ann Male. *Andean Awakening: An Incan Guide to Mystic Peru*.

www.ufocasebook.com/operation mainbrace1952.html.

Darwin, Charles. *On the Origin of Species. Regarding Species of the Oceanic Islands*, 1859: pp. 244–257.

Prothero, D. "Evolution: What Missing Link?" *New Scientist*, February 27, 2008: 35–40.

Morris, Melvin. *Parting Visions*. New York: Villard Books, 1994.

Posnansky, A. *Tihuanacu, The Cradle of American Man*, vols. 1–4. Translated by James Sheaver. New York: 1945 and 1957.

Maloney, M. *UFOs in War Time*. New York: Berkley Books, 2011.

Gee, Henry. *The Accidental Species*. University of Chicago Press, October 2013.

Ann Gibbons. "Human Ancestors Were an Endangered Species," *Science Now*. October 1, 2010.

Rampino, Michael R; Self, Stephen. "Bottleneck in the Human Evolution and the Toba Eruption." *Science* 262 (5142). December 24, 1993.

Watson J.D and Crick F.H. "A Structure of Deoxyribose Nucleic Acid (PFD)." *Nature* 171 (4356): 737-738. 1953.

Smith, Charles H. "Reflections on Wallace." *Nature Magazine*, September 7, 2006.

Flannery, Michael A. "Alfred Russel Wallace: A Heretic's Heretic." *Public Domain Review*.

Schulte, P. et al. *Science* 327 (5970) 1214–1218. March 5, 2010.

HELPFUL WEBSITES

Ancient Aliens (http://www.history.com/shows/ancient-aliens)

www.robertschoch.com

www.ufologie.net/htnjapan86.htn

www.ufocasebook.com

www.ufoevidence.org

www.alienscalpel.com/aliens-and-ufos/
us-president-reagan-extraterrestrial

www.nickpope.net

www.mufon.com

www.dailymail.co.uk/sciencetech/article-1315620/
US-airman-Charles-Halt-UFO-te

www.hyper.net/ufo/rendlesham.htm

www.cufon.org

www.ianridpath.com/ufo/halttape.htm

www.disclosureproject.org

35535587R00155

Made in the USA
Lexington, KY
15 September 2014